fighting FOR THE UNION LABEL

Kenneth C. Wolensky · Nicole H. Wolensky · Robert P. Wolensky

fighting FOR THE UNION LABEL

The Women's
Garment
Industry and the
ILGWU
in Pennsylvania

The Pennsylvania State University Press
University Park, Pennsylvania

LIBRARY OF CONGRESS CATALOGING-IN-PUBLICATION DATA

Wolensky, Kenneth C.
Fighting for the union label : the women's garment industry and the ILGWU in
Pennsylvania / Kenneth C. Wolensky, Nicole H. Wolensky, Robert P. Wolensky.
p.   cm.
Includes index.
ISBN 0-271-02167-5 (cloth : alk. paper)
ISBN 0-271-02168-3 (pbk. : alk. paper)
1. Women's clothing industry—Pennsylvania—History.   2. Clothing trade
Pennsylvania—History.   3. Labor unions—Pennsylvania—History.   4. Interna-
tional Ladies' Garment Workers' Union.   I. Wolensky, Nicole H.   II. Wolensky,
Robert P.   III. Title.

HD9940 .U4 W64 2002
331.4'78187'09748—dc21   2001036463

It is the policy of The Pennsylvania State University Press to use acid-free paper
for the first printing of all clothbound books. Publications on uncoated stock
satisfy the minimum requirements of American National Standard for
Information Sciences—Permanence of Paper for Printed Library Materials,
ANSI Z39.48–1992.

To garment workers,
past, present, and future,
who support families and communities
and whose skills drive the economies
of many nations.

Contents

Acknowledgments

We are grateful to many people for their guidance and assistance in bringing this book to publication. We could not have completed the volume without the participation of numerous individuals familiar with the history of the apparel industry and the ILGWU. They include Min Matheson, Bill Matheson, Betty Matheson Greenberg, Larry Greenberg, Marianne Matheson Kaufman, Helen Sampiero, Sol Hoffman, Lois Hartel, Ralph Reuter, Dorothy Ney, Minnie Caputo, Clem Lyons, Sam Bianco, Tom Mathews, Bob Hostetter, John Justin, Jay Mazur, Martin Morand, Martin and Helen Burger, Helen Weiss, Marilyn Levin, Phyllis Burns, Irwin Solomon, Edgar Romney, Nelson Whittaker Sr., Nelson Whittaker Jr., Celina Whittaker, David Melman, Gail Meyer, Bill Cherkes, Alice Reca, Pearl Novak, Tony D'Angelo, Leo Gutstein, former Pennsylvania governor George Leader, Angelo "Rusty" DePasquale, and Jennie Silverman. Previous research and publications by Gus Tyler, former assistant vice president of the ILGWU, David Dubinsky and Sol "Chick" Chaikin, the union's former presidents, and David Melman of UNITE! proved invaluable as well.

The Research Department of UNITE! and the staffs of Cornell University's Kheel Center Archives in Ithaca, New York, the Pennsylvania State Archives in Harrisburg, the Historical Collections and Labor Archives of The Pennsylvania State University, Temple University's Urban Archives, the New York Public Library, and the Tamiment Institute Library at New York University also deserve our thanks for their patient assistance and guidance with the research that informs this book. The Pennsylvania Department of Labor and Industry's Center for Workforce Information and Analysis proved to be an invaluable resource in providing historical data on employment in the Keystone State's apparel industry. A special thank you as well to Alice Hoffman, whose oral history interviews with Min and Bill Matheson—housed with the Historical Collections and Labor Archives at Penn State—are important supplements to this history.

Acquisition of the many photos that help to inform this history would not have been possible without the generosity of Stephen Lukasik, George Zorgo, Martin Desht, Bernard and Harriet Lurye, Alice Reca, the Kheel Center Archives, the Pennsylvania State Archives, Ace Hoffman Studios, Brown Brothers Inc., and the ILGWU/UNITE!

Our colleagues Brent Glass, Frank Suran, Robert Weible, Linda Shopes, Diane Reed, Marilyn Levin, Michael O'Malley, Curt Miner, and John Zwierzyna at the Pennsylvania Historical and Museum Commission (PHMC), as well as Fred Schied, Daniele Flannery, Simon Bronner, Alison Hirsch, Bill Mahar, Bob Bussell, and Peter Potter at Penn State University, Charlie McCollester of Indiana University of Pennsylvania, and Russ Gibbons, saw the relevance of this work and encouraged its pursuit. Tom Dublin of the State University of New York at Binghamton read and critiqued several drafts, provided data, and supported its publication. Irwin Marcus of Indiana University of Pennsylvania and Robert Janosov of Luzerne County Community College provided important commentary on iterations of this research presented at conferences of the Pennsylvania Historical Association. Barbara Salazar copyedited the manuscript and improved it greatly. Steve Ling, Chester Kulesa, and Hilary Krueger of PHMC's Anthracite Heritage Museum in Scranton, along with George Garner, provided leadership in developing a museum exhibit that illustrates the history discussed in these pages. Joseph Kelly, Laura Clark, and other members of the Pennsylvania Humanities Council graciously sponsored the first author's presentation of this history to numerous historical organizations as part of the 2000–2001 Commonwealth Speakers Program. Richard and Carolyn Williams proofread and commented on the manuscript. The Reverend George Robertson provided moral support in many, many ways. Thanks as well to Courtney Marlaire, professor of sociology at Marquette University, who proofread and commented on portions of our research.

The University of Wisconsin–Stevens Point supported this project by providing two research grants, research assistance in the Center for the

Acknowledgments

Small City, and collegial support in the Department of Sociology. We would like especially to thank Edward J. Miller, Professor of Political Science and Co-director of the Center for the Small City; Justus Paul, Dean of the College of Letters and Science; and Thomas George, Chancellor. A fellowship at the Institute for Research in the Humanities, University of Wisconsin–Madison, allowed Robert Wolensky to work on the Northeastern Pennsylvania Oral History Project, from which many of the data for this study were drawn.

A note of sincere appreciation to our family members, whose encouragement and patience helped to bring this project to fruition. All Wolenskys, they include Abby, Aaron, Molly, Meredith, Jack, Carol, and Rosalie, our mother and grandmother, who once worked in a garment factory. A special note of gratitude goes to Cherie. Without her love and support throughout many years, this book would likely never have come about.

Introduction

Relocating the garment industry was not difficult since it is an industry on wheels. It can be moved overnight because capital investment is low, machines are easily transportable and materials are comparatively light. Clothes are not steel, not copper, not lumber, not brick.
—Sol "Chick" Chaikin, ILGWU president, 1984

For years in many of the old anthracite mining towns the only regular work for women was provided by garment mills. By the late 1980s, many of those enterprises had closed due to low wage foreign imports flooding the American marketplace.
—Richard E. Sharpless, historian, 1995

The union [ILGWU] was good to me. It gave me a lot of opportunities I would never have had. I learned a lot from it. The union was strong. It had leaders who worked hard for the members.
—Rose Liberti, garment worker, 1991

Victor Hugo once said, "There is nothing like a dream to create the future." Indeed, dreamers and visionaries have shaped the American story from Valley Forge to man's first lunar footprints on the Sea of Tranquility in 1969 to the present day. Dreamers have been many and widespread. They have come from all walks of life.

In much of the popular historiography of the United States, entrepreneurs, industrialists, politicians, and other shapers of the public realm and champions of various causes have headed the list of dreamers and visionaries. Social scientists and historians have credited such individuals with certain unique characteristics, such as genius, perseverance, and risk-taking. Success—usually defined as the acquisition of material possessions, various forms of wealth, and power—has been viewed as a just reward for hard work, dedication, and discipline.

In more recent times historians have recognized that among the dreamers and visionaries have been ordinary people who contributed their labor to the American industrial order of the nineteenth and twentieth centuries. In many instances such people uprooted their lives in distant lands, giving up the familiar to search for freedom from subjugation in its various forms. Though their experiences may have come to light only as a result of new ways of looking at history, their accounts tell of another side of the American—and, indeed, the human—story.

Commonly accepted notions regarding this country's past tell us that it was a land of promise where dreams could be realized. Not always, however, did the New World turn out to resemble the mythical shining city on a hill. In this land, too, there were numerous obstacles. Economic, political, and social equity for immigrant working people has been chief among these challenges, both then and now.

The tale of an impoverished late nineteenth-century Italian immigrant inscribed in the main exhibit hall of the museum at Ellis Island best makes the point. The immigrant related that in his native land he was always told that in the United States the streets were paved with gold. Yet when he came here he discovered with his own eyes that the streets weren't paved with gold. In fact, to his amazement, it quickly became apparent that the streets in places like New York weren't paved at all. But even more startling was the revelation that he was the one who was expected to pave them!

The notion remains in popular American historiography that, although he was expected to pave the streets, the Italian immigrant, guided by aspirations for a better life, attained some level of security and comfort through hard work. Yet for many like him the struggle was long and difficult. Indeed, as great as the promise the United States held for the Italian immigrant—and millions of others—he and other working people had to struggle to carve out their place in a society built around democratic industrial capitalism. Indeed, many still struggle as the United States weathers the contemporary storm of downsizing and deindustrialization.

Introduction

The following pages tell a story of people who shared a vision of a better way and the labor organization to which they belonged. Many were immigrants or the children of immigrants who were part of the changing industrial landscape in a place where U.S. industrialization could trace its roots: Pennsylvania's mountainous anthracite or "hard coal" fields. These were people who cut, sewed, stitched, and pressed great swaths of fabrics into clothing for a growing consumer market. They were garment workers. Most were women. The organization to which they paid their allegiance was the International Ladies' Garment Workers' Union (ILGWU).

Anthracite coal was indispensable to the country's growth as a commercial power. During the late nineteenth and early twentieth centuries it was *the* source of energy for countless homes, factories, railroads, and other industrial concerns. By the 1930s, however, its market had declined dramatically. Anthracite workers—mainly men—experienced displacement, unemployment, and underemployment. Though the term would not be coined until later in the twentieth century, when the trend would become a national phenomenon, deindustrialization had begun in Pennsylvania's hard coal fields, probably the first region in the nation to experience such widespread economic decline.[1]

It was by no coincidence that the garment industry gained a foothold in the rugged northern Appalachian hills and valleys where pulling anthracite from the earth was the way most families survived. "Runaway" garment factories—the majority obtaining their work on contract from larger New York manufacturers—set up shop in mining towns where labor was plentiful and apparel unions were scarce. Investors and factory owners were keenly aware that workers—mainly the wives and daughters of unemployed or underemployed mineworkers—came more cheaply and profits more easily when and where economic insecurity was a fact of life. It was therefore advantageous for them to escape or "run away" from garment hubs like Manhattan, where manufacturing costs were higher.

By the 1930s, garment factories were rapidly becoming part of the industrial landscape of the anthracite region. The northern coal fields of the Wyoming and Lackawanna valleys (the only metropolitan areas in the whole hard coal region) were to become a new hub for the manufacture of clothing, mainly for women and children. "Jobbers" secured the work from large New York retailers and manufacturers and sent it to contractors in Scranton, Wilkes-Barre, Pittston, Nanticoke, and numerous other coal towns. Raw materials were shipped to make women's dresses, for example, from New York garment houses to anthracite region factories. Finished products were sold in retail outlets in the eastern United States and elsewhere.

By the late 1930s the ILGWU set out to organize the runaways. The union became concerned about the proliferation of these factories, which harked back to the sweatshops that had permeated Manhattan's Lower East Side at the turn of the century. The contract factories took jobs away from unionized New York workers, lowered wages, and threatened much of what the ILGWU had accomplished since its founding in 1900. Despite concerted efforts, the ILGWU had not made great strides in organizing the Pennsylvania runaways as the 1940s dawned. The industry was growing too quickly, desperate workers were easily swayed by the chance to earn money, and organized crime complicated the picture by owning several shops.

Though runaway factories emerged in many towns in the anthracite region, it was apparent by the 1930s that the Wyoming Valley had become a prime locale. Abundant population, proximity to New York, and access to a transportation network made it attractive to factory owners, as did the lack of any apparel union. In 1937 the ILGWU established locals in several anthracite communities, including Scranton, Wilkes-Barre, Pottsville, and Shamokin. Yet organizing efforts were fitful at best and the growing concentration of the industry in the Wyoming Valley around Wilkes-Barre proved to be a particular challenge. In 1944, however, the situation began to change. In his effort to counter the runaways, the ILGWU's president, David Dubinsky, dispatched Minnie Lurye Matheson to the Wyoming Valley to organize runaway garment factories. In the years to follow Min would play the central role in building a union of over 10,000 members, with major support from her husband, Wilfred (Bill). The following pages tell the story of how the ILGWU in the Wyoming Valley became not only a pay-and-benefits organization but an activist movement that by the 1960s played a vital role in the community; an organization that became perhaps the most widely recognized and respected women's labor organization in the anthracite region and indeed throughout Pennsylvania.

Introduction

This is a case study. It describes how a U.S. labor union used certain tools, techniques, and strategies to organize a large portion of the industry that sought to escape it. It explores how unionists—some of whom were dreamers and visionaries, others pragmatists and technicians—built an infrastructure and exposed working people to a culture in which knowledge, activism, and notions of community were highly valued. It examines how a labor organization concerned itself not only with conditions on the shop floor but with larger issues of economics and politics and their impact on the community. And as the last two chapters explain, the runaway garment industry is not merely an icon from the past, for it is apparent that apparel manufacturers have continued to migrate for the same reasons that they moved to Pennsylvania and its anthracite region more than a half-century ago.

This study also offers a rare look at garment workers and the ILGWU outside of New York, which in most of the annals of the apparel industry and its unions has remained at the center of its memory.[2] Sol Hoffman, who dedicated his forty-plus-year career in the labor movement to the ILGWU, which he served as vice president, addressed the novelty of examining its history in locales other than New York:

> When they write the history of the union, I don't know how much of this [non–New York] story they will have. Keep in mind that Pennsylvania, or anything outside of New York, is not seen as the history of the union. But I don't know of anyone outside of the New York area who has had their story dwelled upon to any extent. Obviously, most of the members and the founders of the union were from New York, that's true. But there were a lot of heroines and heroes out there in the hinterlands who you never hear of; who are never written about. Min Matheson was outstanding among them.[3]

In sum, this history is a microcosm of the rise and fall of the women's apparel industry and its preeminent labor union during the latter two-thirds of the twentieth century.

The most significant primary source of information for the study was the 325-person Northeastern Pennsylvania Oral History Project, which we directed. Begun in the early 1980s, the collection is housed at the Center for the Small City, University of Wisconsin–Stevens Point, and includes the taped memoirs of a diverse selection of people who have resided in or possess direct knowledge about this region of the Keystone State. For this study we have made extensive use of interviews with garment factory workers and owners, union leaders and activists (in both Pennsylvania and New York), public officials and politicians knowledgeable about the industry and the ILGWU, Min Matheson,[4] and others with related knowledge.[5]

The oral history collection of the Historical Collections and Labor Archives at the Paterno Library of the Pennsylvania State University at University Park provides another valuable primary source. In the early to mid-1980s, Alice Hoffman, then a faculty member in Penn State's Labor Studies program, conducted several interviews with Min and Bill Matheson. In these interviews Min details many of the same historical recollections as in her Northeastern Pennsylvania Oral History Project interviews, thus validating her memories of people, places, events, and circumstances.[6]

Additional primary sources include ILGWU convention reports; papers of the ILGWU Education Department; David Dubinsky, Charles Zimmerman, Gus Tyler, David Gingold, and other ILGWU leaders; and oral histories in the ILGWU's official archive, housed at Cornell University's Kheel Center for Labor Management Documentation, School of Industrial and Labor Relations, Ithaca, New York. Other primary sources and references include the papers of Governor George M. Leader and records of the Pennsylvania Department of Labor and Industry, both housed at the Pennsylvania State Archives in Harrisburg; records of the Philadelphia Joint Board of the ILGWU, housed at Temple University's Urban Archives; and the papers of Fannia Cohn, housed at the New York Public Library.

The study also drew upon materials in *Justice*, the ILGWU's official newspaper; *Needlepoint*, the newsletter of the ILGWU's Wyoming Valley District; Pennsylvania Crime Commission studies and reports; various uncatalogued ILGWU documents, scrapbooks, and publications housed in the library of the Wyoming Valley District or lent by union members and leaders; newspaper accounts; and the writings of individuals once active in the union, such as presidents David Dubinsky and Sol "Chick" Chaikin, Gus Tyler, Leon Stein, David Gingold, Mark Starr, and Fannia Cohn. Secondary sources have been examined to highlight the history and social and economic contexts of Pennsylvania's anthracite region. All primary and secondary sources are named in the notes.

Chapter 1 highlights the transition of apparel manufacturing from "homeworking" to the factory; discusses the evolution of the ILGWU and some of the early struggles of garment workers; and describes the genesis of runaway garment factories. It also highlights Pennsylvania's experiences with industrial sweatshops. The chapter concludes by introducing Pennsylvania's anthracite coal region and discusses the rise of the apparel industry in the area.

Chapter 2 explains the ILGWU's early efforts to counter the growth of the runaways. It introduces Min Lurye Matheson and Bill Matheson and describes Min's discoveries in Wyoming Valley garment factories in the mid-1940s. This chapter discusses the influence that Min's father—a Chicago cigarmaker and union organizer—had on her as a youngster and highlights her background as a young woman among labor activists in New York and in the ILGWU.

Chapter 3 describes strategizing, organizing, and building momentum for the ILGWU in the Wyoming Valley. Narrators describe encounters with organized criminal elements that infiltrated or influenced some garment factories, particularly in Pittston, a coal town at the northern end of the valley. The chapter details how, despite some momentum, the murder of Min's brother, William Lurye, increased the very real danger she and her colleagues faced.

Chapter 4 examines how the ILGWU worked to establish an infrastructure to provide for the human needs of garment workers. Rooted in the union's culture of education and activism—a culture that transcended individualism and associated members with one another and with the larger organization to which they belonged—the initial components included a routinely published newsletter named *Needlepoint;* a union-operated health care center, the only one of its kind in the anthracite region; and a popular and successful chorus.

Chapter 5 explains how the union's infrastructure was expanded to include educational programs, community participation, and political activism so that garment workers might acquire knowledge about the practicalities of unionism as well as their role in U.S. industrial capitalism. One clear example of the union's community-mindedness and political involvement was its work to improve the dismal economic situation in the anthracite region. Espousal of public policies to stimulate economic development demonstrates that the union had moved well beyond its bread-and-butter concerns.

Chapter 6 scrutinizes the 1958 general dress strike and its impact on the Wyoming Valley District, as well as an extended strike against a Pittston manufacturer backed by one of the nation's most infamous crime figures, Rosario Bufalino. The chapter also describes a successful campaign by resolute women workers to drive the gangster Tommy Luchese (a.k.a. Three Finger Brown) out of the ownership of a factory. These events illustrate that despite its gains in a variety of arenas, the ILGWU continued to face formidable obstacles, including organized crime and a lack of wage parity with apparel workers in New York.

In Chapter 7 Min Matheson departs the Wyoming Valley in 1963 and leaves behind a vigorous and growing union. Just a few years before her relocation, however, few could have foreseen that the importation of Japanese-made scarves would one day translate into the near demise of an entire U.S. industry. Not long after Min Matheson relocated to New York, a new reality became obvious to garment workers and the ILGWU: imports flooded U.S. markets, manufacturers built factories in the South and later overseas to take advantage of cheap labor, and government policy would do little to protect domestic jobs. Obviously, the runaway garment factory is by no means a phenomenon of a bygone era.

In the early 1970s the ILGWU began a long decline. Deindustrialization was a reality for apparel workers. As Chapter 7 demonstrates, what was happening to the garment industry in Pennsylvania generally and in the anthracite region specifically would, once again, represent a microcosm of a larger problem. The union's struggle with Leslie Fay, Inc., one of the nation's premier women's clothing manufacturers, came to signify a desperate stand against the power of international capitalism and the lure of the global economy.

The Epilogue discusses ways in which this hundred-year-old labor organization has responded to the globalization trend. During the past century the ILGWU's challenges have shifted from Manhattan's Lower East Side to the anthracite region of Pennsylvania, from there to the South and elsewhere in the United States, then across the vastness of the oceans and in some cases back to the United States' large urban centers, where sweatshops have reappeared.

As Min and Bill Matheson began their work in the Wyoming Valley, they hardly could have thought that what the ILGWU characterized then as the "out-of-town problem" would be more accurately called an "out-of-country" problem a half-century later. Today the apparel industry continues to migrate for the same reasons that it escaped from New York to Pennsylvania's anthracite region in the 1930s and 1940s. A half-century ago, however, the ILGWU had the capability, in the United States at least, to dispatch organizers to challenge runaway factory owners and negotiate contracts for workers. Today the challenges are the same but the economic geography is much different. The runaway garment factory is now a global phenomenon.

In the following pages we write about and interpret the past in an effort to understand how the past influences the present and provides insight for the future. We also attempt to explicate what the past has meant to those who lived it. To illustrate this point and to introduce this story, we turn to Marianne Kaufman, the Mathesons' daughter, who provided her assessment of the work of the ladies' garment workers' union in the Wyoming Valley by offering the following allegory: "On a bright summer day two people were walking along an ocean beach where hundreds of starfish had washed up on shore. One of them picked up a starfish and threw it back into the ocean. The other person said, 'What difference did that make?' to which the first person responded, 'It made a difference in the life of that starfish.' "[7]

I

An Industry, a Union, and Runaway Garment Factories

"Runaway" shops . . . are established principally because of cheap labor.

—*Report and Proceedings of the Tenth Convention of the* ILGWU, 1922

Sweatshops are the outstanding evil of industrial life in this state.

—Governor Gifford Pinchot, 1927

There are some [runaway garment factories] in hiding but not for long.

We [the ILGWU] will search them out like bloodhounds.

—David Dubinsky, 1949

As the 1960s drew to a close, the apparel industry employed nearly 180,000 Pennsylvanians in 2,000 factories. Eighty thousand of these workers were members of the International Ladies' Garment Workers' Union (ILGWU), about two-thirds of them women who produced clothing for women and children. Their places of employment were concentrated in five main geographic areas of Penn's Woods. First was Philadelphia and its immediate environs; second, the Lehigh Valley, extending west to Reading and Berks County; next, central Pennsylvania, including the Lebanon Valley, the greater Harrisburg area, and York and Lancaster; fourth, western areas of the state, including Johnstown, Altoona, and Pittsburgh; fifth and finally, the garment industry was highly concentrated in the Appalachian mountains and valleys of the northeastern corner of Pennsylvania, otherwise known as the anthracite region.

How garment making grew as an industry and, temporarily at least, prospered in Pennsylvania and its anthracite coal fields can be traced to its genesis in Manhattan, where sweatshops flourished, its workers began to organize, and the industry began its history of migration in the late nineteenth and early twentieth centuries.

Garment-Making Industrializes

As the end of the nineteenth century drew near, the manufacture of clothing in the United States shifted from small artisan shops, individual homes, and immigrant-occupied tenements in urban industrial areas such as New York's Lower East Side to a growing number of factories. Though the roots of factory-based apparel making can be traced to the 1840s, several factors spawned its expansion at century's end.

The advent of the foot-powered sewing machine in 1846 and the mechanical cutting knife in 1876, as well as the introduction of steam power by the 1890s, meant that large quantities of fabric could be made into finished clothing with greater efficiency. Technology transformed and specialized labor while speeding production to meet the growing consumer demand for ready-made clothing. Advances in transporting raw materials to manufacturers and finished goods to consumers opened new markets. And concentrations of immigrants eager for work provided the necessary workforce.

In 1880 the typical New York garment-making establishment was owned by a German Jewish clothing retailer who employed a small number of people. Most of the workers who made clothing for women and children—the focus of this study—were female. In the years to follow tens of thousands of Jews emigrated from Russia and Eastern Europe in flight from pogroms, religious persecution, and political discrimination.[1] Many possessed skills as tailors and seamstresses and entered the women's clothing industry en masse, to be joined by Italian and other immigrants. By the dawn of the twentieth century the influx of these immigrant groups had a significant impact on apparel making.[2]

As factories became common in the U.S. industrial landscape, so too they became an important part of garment making early in the new century. Many garment workers were employed in so-called inside shops, establishments owned by a manufacturer, who purchased raw materials, planned production, and hired immigrant labor to cut, sew, press, and finish fabric into consumer products. The manufacturer arranged for their sale in the marketplace. The main branches of the women's apparel industry produced dresses, cloaks and suits, corsets and brassieres, other undergarments, neckwear, rainwear, and infants' and children's wear. Although some large factories emerged in East Coast cities such as New York, Boston, Philadelphia, and Baltimore, most garment-making establishments remained relatively small. The average factory employed twenty to thirty people in 1899, a number that would not significantly change through the Great Depression. Though women's apparel making would expand to cities from coast to coast—Philadelphia, Cleveland, Chicago, St. Louis, Los Angeles, San Francisco—and to outlying areas, Manhattan would remain a prime locale throughout the twentieth century.

Unionization began in the same city in 1879. The Knights of Labor established a short-lived workers' association, which was superseded in 1883 by the Dress and Cloak Makers' Union and the Gotham Knife Cutters' Association of New York and Vicinity. Unions were also formed in Toledo, Baltimore, and Philadelphia. Worker-led protests against low wages, long hours, and deplorable working conditions were common. The entire cloak trade workforce in New York walked off the job in August 1885 demanding higher wages and a shorter workday. In 1886 workers struck to protest the practice of contracting work to small nonunion producers who paid extremely low wages. Several new unions participated in the strike, including the Independent Cloak Operators' Union and the Independent Cloak Pressers' Union. The following year thirty walkouts were reported in the New York area over similar issues. Numerous garment strikes hit New York and Philadelphia in 1888 as workers continued to demand improved working conditions and higher wages. These disputes were typically set-

A tenement sweatshop on the Lower East Side of New York, early twentieth century. (Courtesy of Kheel Center Archives, Cornell University.)

tled with limited concessions to the workers. The conclusion of a strike usually meant dissolution of the labor organization that spawned it. In the waning years of the 1880s, for example, at least a half-dozen unions emerged and quickly dissolved in New York alone.

By the 1890s, the socialist United Hebrew Trades supported garment unionization by fostering the creation of the Operators' and Cloak Makers' Union No. 1 in Manhattan. Affiliated organizations emerged in Chicago, Boston, Baltimore, and Philadelphia. After a major strike in New York in the spring of 1890, employers recognized the union. By the end of the year the Operators' and Cloak Makers' Union No. 1 reported a membership of over 7,000. By 1892, however, the union was in disarray, in part because of the

An Industry, a Union, and Runaway Garment Factories

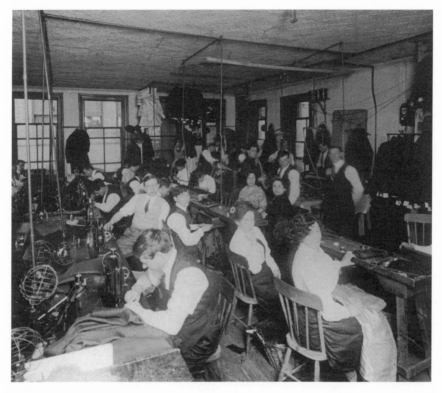

A tenement sweatshop on the Lower East Side of New York, early twentieth century. (Courtesy of Kheel Center Archives, Cornell University.)

imprisonment of its controversial and outspoken leader, Joseph Barondess, upon his conviction for arson and assault. Though Barondess was later released and pardoned, some union members advocated his banishment.

The tumultuous drive for unionization continued in 1892 with the formation of the International Cloak Makers' Union of America, headquartered in New York, where Barondess managed its metropolitan branch. The union had affiliates in Boston, Baltimore, Chicago, and Philadelphia, and joined the American Federation of Labor (AFL). Barondess played a leading role in the formation of the organization. The new union also set out to organize men's tailors and clothing makers, the traditional domain of the more conservative AFL-affiliated United Garment Workers Union.

Over the next few years the pattern of union formation and dissolution in women's apparel continued, as did infighting among socialist, anarchist, and conservative forces. In June 1900, representatives from the Cloakmakers Protective Union of Philadelphia, United Cloak Pressers of Philadelphia, Cloakmakers Union of Baltimore, United Brotherhood of Cloakmakers of New York and Vicinity, Newark Cloakmakers Union, and the Shirtmakers Union of New York met at Manhattan's Labor Lyceum, at 64 East Fourth Street, to discuss ways to stabilize the rancorous relations between apparel workers and manufacturers. On the agenda was the idea of creating a single labor union to represent women's and children's garment workers. The conclave resulted in the formation of the ILGWU. Though the new union was centered in New York, its mission was to improve wages and working conditions in women's garment factories throughout the United States and Canada. The ILGWU affiliated with the American Federation of Labor.[3]

A New Union with Old Problems

The new union's first decade was marked by drives to organize workers, raise their wages, and win recognition. Garment manufacturers—in part to avoid the ILGWU and its demands for higher wages and better working conditions—expanded on a practice that had roots in the nineteenth century: contracting to outside shops. Under this arrangement, manufacturers and jobbers purchased fabrics, designed new styles, then contracted out for the manufacture of the garments.[4] Small highly competitive contractors bid on the orders. They invariably paid low wages and demanded long hours from employees, who worked at a frantic pace in cramped and unsafe conditions to fill orders. Indeed, some manufacturers combined both the "inside" and "outside" systems, producing some apparel in establishments of their own and securing contractors to fill selected orders. Even while mass production of clothing was in its infancy in the nineteenth century, the manufacturer/jobber-contractor system expanded as producers sought to enhance profits and reduce turnaround time. Early contractors included homeworkers, tenement-based production workers, and contract shops located in New York's Coney Island. With its notoriously low wages and poor working conditions, contracting gave rise to the industrial sweatshop so often associated with garment making on Manhattan's Lower East Side. Though this system threatened unionization, two significant strikes and a tragic fire brought public recognition to the struggles of garment workers. The episodes would also prove significant in advancing recognition of the ILGWU.

In 1909 20,000 shirtwaist makers—mainly women—struck New York employers, among them the Triangle Shirtwaist Company. With support of the Women's Trade Union League, workers protested unsafe working conditions, low wages, long hours, and the imposition of employer-imposed "taxes" for electricity, sewing needles, and chairs. The ILGWU's shirtwaist makers' affiliate, Local 25, spearheaded the strike, dubbed the "Uprising of 20,000." Over 600 employers were affected. When the strike was settled, about half of the employers agreed to a shortened workweek (52 hours), elimination of discrimination against union members in hiring, an end to all "taxes," four annual paid holidays, and union recognition. Many holdout employers eventually granted some of the workers' demands, but the ILGWU failed to gain full recognition.

The Uprising was followed in 1910 by the "Great Revolt" of 60,000 New York cloakmakers. The grievances that brought these workers to the picket line were similar to those that sparked the Uprising. Employers responded to the strike by forming the Cloak, Suit, and Skirt Manufacturers Protective Association and issuing public statements claiming that workers were treated fairly. Louis Brandeis, a Boston lawyer and later U.S. Supreme Court justice, mediated a settlement of the dispute by creating the historic Protocol of Peace.

While the Protocol was in its infancy, tragedy struck at the Triangle Shirtwaist Company in March 1911; the event would prove pivotal in fostering the union's cause. Located at New York's Washington Square, Triangle employed hundreds in a cramped, unsafe, multistory building where workers logged days of twelve hours or more. On March 25, 1911, during the daytime shift, a fire of uncertain origin broke out on the middle floors of the building. Fueled by an abundance of cloth and other material, flames quickly spread. Frantic workers screamed for help and gasped for

The Uprising of 20,000 New York shirtwaist makers, 1909. (Courtesy of Kheel Center Archives, Cornell University.)

air at upper-story windows as flames billowed around them. Rather than succumb to the inferno, some leaped from windows as high as eight stories. Others could do little more than await firefighters, hoping that the flames could be tempered and an escape route discovered. For many, death proved the only deliverance. Sixty-two jumped to their deaths on the sidewalk below, some splitting safety nets held by firefighters to break their fall. For those who remained in the building, the outcome was equally

Victims of the Triangle Shirtwaist factory fire. (Courtesy of Brown Brothers, Inc.)

bleak, as exits were either locked or cluttered with equipment, cloth, and refuse. In all, 146 garment workers died.[5]

Strikes and the Triangle tragedy set the stage for Brandeis's labor–management relations experiment. The Protocol of Peace established several important precedents by creating a Committee of Grievances and a Board of Arbitration to mediate and arbitrate employer–union disputes and worker grievances; a Joint Board of Sanitary Control to oversee factory and employee health and safety; and a preferential union shop in which hiring was limited to union members. The Protocol had no expiration date, and Brandeis envisioned it as a way to maintain amicable relations in a highly unstable industry. While it did not immediately resolve all disputes, it

was notable as an early attempt to establish a platform for modern industrial and labor relations. In a few short years its status would be jeopard-
ized by enforcement problems and by the growing practice of contracting.
After numerous complaints to the New York Cloak Joint Board—the ILGWU
affiliate that represented workers in various trades associated with the
manufacture of cloaks and suits—hapless efforts were made in 1913 to
amend the Protocol so that wages paid in contract shops were on a par
with those in inside shops.[6]

While the industry and the ILGWU remained centered in New York during the early years of the Protocol, the union expanded in other eastern
cities. Philadelphia, Pennsylvania's early twentieth-century garment manufacturing hub, was one such locale. A 1912 census by the ILGWU's Philadelphia Cloakmakers' affiliate, Local 2, reported a membership of 1,000. The
ILGWU viewed the city as "unfavorable for successful trade unions."[7]
Employers defeated a citywide Cloakmakers' strike in 1914 for union
recognition and other concessions. By 1915, the union reported that nearly
6,000 women toiled in the city's dress shops without representation. That
same year the ILGWU called a general strike of shirtwaist- and dressmakers
and formed Dressmakers' Local 15, enrolling 5,000 workers and winning
wage and hour concessions. Despite some success, the ILGWU would not
win recognition in Philadelphia until a general dress industry strike in
1933.

By the end of World War I the ILGWU had a membership of about 100,000
nationwide. During the following decade the union was besieged by internal struggles with Communists, who, as the ILGWU historian Gus Tyler
notes, followed orders from the Soviet Union to take control of organized
labor as part of a crusade to spread communism around the globe.[8] In 1920
William Z. Foster—who was to become perhaps the best known communist in the United States—formed the Trade Union Educational League
(TUEL) to coordinate the work of leftist activists in the U.S. labor movement. TUEL policy was influenced by Moscow and the U.S. Communist
Party. It advocated class struggle, international unification of industrial
workers, mass organization across industries, creation of a labor party, U.S.
recognition of the Soviet state, and destruction of capitalism. Radicals
within the ILGWU, influenced by the TUEL, sought to usurp control through

the Party's strategy of "boring from within." Communists organized workers' councils in ILGWU locals. Although the ILGWU banned the councils in 1921, the left wing—led by Louis Hyman of the cloakmakers and Charles (Sasha) Zimmerman of the dressmakers—expanded its influence in the ILGWU's Local 22, the Dressmakers' Union; Local 25, the Shirtwaist Makers' Union; and Local 9, a cloakmakers' affiliate.

By 1924 the left elected a majority to the executive boards of key New York locals, including 9 and 22, and refused to recognize the legitimacy of the ILGWU's president, Morris Sigman. Sigman, a Socialist who rejected Communist dogma, expelled some TUEL supporters and refused to allow others to hold union office. To assert greater authority over the union he also merged both the Cloak Joint Board and the Dress Joint Board into a single entity.[9] The move would eventually backfire, as it paved the way for the left to effectively control the new joint board and establish a dual union movement.[10] According to Jennie Silverman, who worked as a dressmaker and associated with the Communists:

> The union had been split. The ILGWU had been split by the Communists who were members of the union. The [Bolshevik] revolution was still young and was a source of inspiration to so many. They were always on the picket line and they took their responsibilities so seriously. The Comintern gave an order, "Take over the unions or break them," which became known as the "Rule" or "Ruling Policy." And that's exactly what they did. The Communists took over. They were always very cocky. The future was always theirs. They brainwashed their people by making them paranoid. They absolutely preached paranoia. [Their message was] "Everybody's against you—the teachers, professors, shop owners, union people, intellectuals. They're all in the employ of the capitalist class, if not directly, then indirectly." So I joined [the Communists] when I was sixteen. What the hell did I know?[11]

Conflict between radical and moderate forces peaked with two significant events. Among the Communists' most vociferous arguments was that the union lacked democracy, as exemplified by the fact that union locals were permitted to send only five delegates to ILGWU conventions regardless of their size. They argued for proportional representation: the larger the membership of a local, the more delegates sent to a convention. Morris Sigman agreed to call a special convention in Philadelphia in the fall of 1925 to address several matters, including proportional representation. Though the intent was to arrive at a peaceful resolution of contentious issues, the convention proved to be raucous. Delegates remained divided over the representation issue and an earlier decision of the General Executive Board (GEB) to suspend members who affiliated with the Communist fringe was upheld. Hyman led the radicals in a massive walkout. Contentions remained unresolved.

When the contract between the Joint Board and cloak manufacturers was renewed in early 1926, Communists led a strike—unauthorized by the GEB—of nearly 40,000 New York–based cloakmakers. Among the many issues that angered radicals and the Communist Party apparatchiks who ordered the strike was that the ILGWU had little control over work that was contracted to small nonunion factories. The shutdown lasted twenty-eight weeks at a total cost to the ILGWU treasury of $3 million. In an effort to bring the strike to a conclusion, President Sigman dissolved the Joint Board, whose leftist leanings had spawned the strike. He established a new board and encouraged strike-weary, hungry, and penniless workers to affiliate with the organization governed by the union's more moderate forces. A majority of the workers came back into the ILGWU fold and Sigman successfully negotiated a contract with the employers, although it lacked significant restrictions on contracting. Meanwhile, Communist policy makers instructed radicals to walk away from conservative U.S. labor unions and align themselves with the newly emerged Trade Union Unity League, which intended to give birth to a revolutionary movement. In the needle trades disenchanted leftists joined the Needle Trades Workers' Industrial Union. Not all of the Communists, however, followed this path. Hyman and Zimmerman rejected the move and remained with the ILGWU, becoming stalwart anti-Stalinists.

In 1928 Benjamin Schlesinger would once again assume the ILGWU's presidency (he had served in the post on two previous occasions: 1903–1904 and 1914–1923). David Dubinsky, the head of Local 10, representing garment cutters, assumed the post of secretary-treasurer. By that time the Communists had been expelled from ILGWU locals in New York, Boston, Philadelphia, and Baltimore. Yet the impact of their actions and of the 1926 strike endured. The union was bankrupt. Membership dwindled to 60,000. Moreover, New York members were increasingly threatened by unemployment and underemployment as the practice of contracting garment production to nonunion factories outside the city expanded.[12]

Although Communists argued that the ILGWU did little to mitigate the practice of contracting, the General Executive Board voted at its 1922 Cleveland convention to levy a $4 per capita tax on members as part of an effort to survey and organize contractors. Three issues exacerbated the union's concern. First, garment factories were emerging in areas on the outskirts of and remote from the New York metropolis. Dubbed "runaway" and "out-of-town" (i.e., outside Manhattan) shops, these factories were increasingly difficult for union organizers to police, as the ILGWU had yet to establish a significant presence outside the large cities. Second, because these factories were nonunion, employees were paid less and worked in far worse conditions than those employed in union-ized shops. As a result, the contract factories threatened union wages, piece rates, and working conditions that the ILGWU had negotiated for its members.

Third, manufacturers and jobbers increasingly resorted to contractors, especially during periods of peak production. Because most people bought new clothing as the seasons changed, demand for certain types of clothing was greater in some months than in others. Lightweight warm-weather dresses, for example, were in high demand in spring and early summer. To supply the market and avoid expanding their own production capacity—a sometimes unwise business decision, given inevitable fluctuations in demand—manufacturers used contractors for certain types of apparel when demand was greatest. The arrangement proved convenient. Manufacturers could supply the market while avoiding hiring costly personnel. Contractors were quick to hire workers and set up shop to secure a speedy profit.

The vote at the 1922 convention to levy a $4 per capita assessment led to the formation of the Eastern Out-of-Town Organizing Department. Based in New York, the department was charged with organizing the contractors that were sprouting in areas removed from the city. A union survey conducted before the convention estimated that some 270 shops employed 17,000 women's apparel workers in over 50 small towns and villages in New York, New Jersey, Connecticut, and Pennsylvania. Though it could not be determined with any certainty whether all of these shops were indeed runaway contractors, the ILGWU leadership nevertheless argued that organizing these shops would help place their wages and working conditions on a level with those of the union shops in Manhattan. By the mid-1920s the Eastern Out-of-Town Department enlisted 2,500 members organized in 29 locals in New Jersey, Long Island, and Connecticut.[13]

By the time David Dubinsky assumed the presidency of the ILGWU in 1932, the union's future appeared grim. The onset of the Great Depression had caused a significant industrial slowdown. ILGWU membership dwindled to fewer than 25,000 and the union's financial situation remained precarious. It was also clear that the problem of the out-of-town shops had grown to major proportions. As the nation's economy stalled, competition among jobbers put tremendous pressure on contractors, who were played against one another for even the smallest margins. Contractors responded by ignoring union agreements, paying below-market wages, breaking the union altogether, and seeking the cheapest labor possible by fleeing in ever greater numbers.[14] Jobbers and manufacturers, ever concerned about profit margins, encouraged this new arrangement so enthusiastically that they became known as "jobber barons." The ILGWU explained at its 1932 convention in Philadelphia:

> The out-of-town problem is not a separate problem by itself but is closely tied up with the situation in New York both in the cloak and dress industries. The number of shops in the suburban territory grows when it becomes profitable for the New York cloak and dress jobbers to encourage contractors to move their shops or open a few shops out of town. There are, according to figures obtained by us, not less than 150 dress shops located within a radius of seventy miles from New York employing several thousand workers at unbelievably low wages.[15]

It was relatively easy to establish a runaway shop. Historically, the manufacture of apparel required little initial capital. All that was typically needed to set up operations was a locale, a few sewing machines, and a contract. Once all three were in place, labor could be hired. Labor accounted for the largest proportion of ongoing capital outlay. To secure work, a contractor engaged in bidding wars with competitors. Outbidding the opposition meant driving down wages, because costs for other capital outlays, such as rent and equipment, remained relatively constant.

As Sol Chaikin, later president of the ILGWU (1975–86), explained, the apparel industry and in particular the contracting system represented the purest form of free enterprise, market-driven capitalism. Poorly financed entrepreneurs had no choice but to underbid one another to stay in business. The only alternative was to go out of business (which many did). Loss of a contractor would be of no concern to a jobber—contractors were easily found. Employees had little choice but to work long hours for poor compensation as their wages were squeezed in the bidding process. The only alternative was unemployment. According to Chaikin, because of the ease of entry into apparel manufacturing—from the beginnings of the homeworking system to the advent of mass production to the growth of the overseas industry in the closing decades of the twentieth century—apparel making has indeed been an industry on wheels. Further, Chaikin explained: "Relocating the garment industry was not difficult. It can be moved overnight because capital investment is low, machines are easily transportable and materials are comparatively light. Clothes are not steel, not copper, not lumber, not cement, not brick."[16]

As the contracting system expanded during the 1920s and 1930s, out-of-town shops received nearly all their work from New York jobbers. An order for women's dresses, for example, often meant that the pattern might be cut in New York, sent to a contractor in Connecticut, New Jersey, or Pennsylvania for assembly and finishing, and shipped back to New York for retail sale. The exodus of the contractors to distant areas was well known to those who followed industry trends. According to Frances Perkins, labor secretary under President Franklin D. Roosevelt:

Since he [the contractor] cannot hope to meet union conditions or the requirements of labor law, he goes to some outlying suburb where garment factories are not a feature of the local picture and where state inspectors are not on the lookout for him. Or perhaps he goes to a nearby state—New Jersey, Connecticut, Pennsylvania, Massachusetts—where he believes labor laws are less stringent or that he will escape attention. The goods he makes up are probably cut in a city shop and "bootlegged" to him by truck. His operations are minutely subdivided so that they can be quickly learned and require little skill. His work force is made up of daughters and wives of local wage earners who have been out of work for months or even years and whose family situation is desperate. The boss sets the wage rates, figures the pay slips, determines the hours of work. His reply to complaints is, "Quit if you don't like it." In the *runaway* shop conditions are usually far below standard and the picture of such a plant is a look back to the sweatshops that horrified case-workers and visiting nurses at the turn of the century.[17]

Likening them to nothing more than sweatshops, the *Christian Science Monitor* referred to the out-of-town contract factories as "gypsy manufacturers" who cared about little else than quick profits: "The gypsy contractor does business on a shoestring, choosing for his location places outside the inspection rounds or in a state where labor laws are laxly enforced. In New York State he prefers Long Island, Staten Island and Westchester County, or he goes to Pennsylvania, Massachusetts, Connecticut, New Jersey, trucking his stuff into the retail center at night."[18]

Indeed, the industrial sweatshop was nothing new to Pennsylvania. In the 1890s the Commonwealth's Office of the Factory Inspector (which in 1913 became the Department of Labor and Industry) was statutorily mandated to examine and report on working conditions in the Keystone State's garment industry. Inspectors routinely uncovered high rates of worker injury, child labor, unsafe workplaces, and other sweatshop problems in tenements and factories. In 1895 Governor Robert Pattison signed the Tenement House Law—dubbed the Sweatshop Law—to outlaw the manufacture of clothing in tenements except for items made for personal use. In effect the law helped to spawn factory-based apparel making while broadening the ability of the factory inspector to police conditions in the industry. Though the statute eliminated tenement sweatshops, conditions in the factories were not very different. A study of Pittsburgh's apparel industry found contractors' shops—some of which had received work from Cleveland-based jobbers—that clearly fitted the sweatshop description:

> Here the day begins at seven o'clock. The lunch hour is a casual interval, often omitted. The close of the day comes sometimes at six, sometimes at nine, sometimes at ten; in fact, whenever the work is finished. Whereas none of the garment factories has a working day of more than nine and a half hours, the contract shops have a working day of ten to fourteen hours. Whereas the garment factories employ English speaking, frequently native-born girls, who by their unions have secured for many of their number a living wage, the contract shops employ immigrant labor at rates which often enable them to underbid factory production. The factory equipment may be unsuited to the filling of special orders.[19]

Such working conditions attracted attention from public policy makers in the 1920s during Governor Gifford Pinchot's first nonconcurrent term of office (1923–27). Previous attempts by Harrisburg to regulate the work of women and children in textile, apparel, and other industries included the 1913 Women's Labor Law, limiting total weekly work hours to 54, and the landmark 1915 Child Labor Law, which set a minimum working age of 14 and forbade nightwork by youth. Arguing that these measures were not enough and that sweatshops were "the outstanding evil of industrial life of this State," Governor Pinchot authorized the creation of a Bureau of Women and Children in the Department of Labor and Industry to study the problem further and recommend legislative and regulatory amendments.[20]

During the Great Depression the Department of Labor and Industry, prompted by a series of strikes by women and children in Allentown-based shirt factories, researched sweatshops and reported on wages, working conditions, and child labor. In its 1933 survey of 10,000 female and child apparel workers in Allentown, Doylestown, Philadelphia, and Shamokin the Department discovered that

> pittance wages predominated. For a full week's work more than seventy-five per cent received less than \$5; nearly one-half earned less than \$3 and more than twenty per cent received less than \$2. The median average weekly earnings of \$3.31 for 14 and 15 year old children in the clothing industry in 1932 was a decrease of sixty per cent from the \$8.38 median for the same age group in the same industry in 1926. By 1933, when another survey of the clothing industry was made, the median had declined farther to \$2.76.[21]

Department inspectors also discovered that clothing was manufactured in people's homes under contract. Women and children constituted the majority of this workforce.

The agency conducted a follow-up survey in 1934 after passage of the New Deal's National Recovery Act (NRA), which banned some types of homework, mandated a 40-hour week, established a 16-year minimum working age, and set a minimum wage. While inspectors found homework on the decline and instances of shorter hours and increased wages, they reported inconsistent and sporadic NRA compliance and the persistence of the manufacturer/jobber-contractor system.

As Washington lawmakers debated the merits of enacting Roosevelt's NRA, Governor Pinchot, during his second nonconcurrent term in office (1931–35), persuaded the state's General Assembly to establish a committee to investigate sweatshops and develop legislative recommendations. Advocating reform was Cornelia Bryce Pinchot, the governor's wife, a social activist committed to Theodore Roosevelt–style progressivism, who had called public attention to women's labor issues as early as the Triangle Shirtwaist factory fire. The General Assembly followed the advice of the governor and his wife to investigate sweatshops; but Pinchot later criticized legislators for their failure to grant the Department of Labor and Industry greater enforcement powers:

> Cornelia Bryce Pinchot first called State and Nation-wide attention to the terrible conditions of Pennsylvania's sweatshops by her own

investigations of the needle trades. Her work brought on a legislative investigation which pointed out the crying need for minimum wage and maximum hour laws and the complete elimination of child labor. Many women were found to be working 50 and 54 hours a week for one, two, and three dollars and large numbers of fourteen and fifteen year old children were employed. The General Assembly failed to act on the report of its own committee.[22]

In 1934 Pennsylvanians elected George Earle to the governor's office. A reform-minded New Deal Democrat, Earle signed several sweeping workplace amelioration measures as part of Pennsylvania's "Little New Deal." The Industrial Homework Law tightened restrictions and, in many cases, prohibited employers from hiring people to engage in production work in their own homes; the Minimum Fair Wage Law for Women and Minors established a minimum wage for women and children; the Pennsylvania Labor Relations Act guaranteed employees the right to organize and bargain collectively and created the Pennsylvania Labor Relations Board to protect organizing efforts; and the Pennsylvania Labor Mediation Act conferred the power to mediate and arbitrate labor disputes on the Department of Labor and Industry. These laws either mirrored or filled in gaps not addressed by federal New Deal measures and affected workers in manufacturing industries, including apparel.[23]

While state government was taking action to regulate activities in the workplace more closely, sweatshops continued to flourish. Indeed, to the ILGWU the growing out-of-town problem had created a novel twist by the mid-1930s. During the first few years of David Dubinsky's three-decade-plus career as the union's president (1932–66), more and more garment factories sprouted in the Keystone State's anthracite region.

Pennsylvania's Anthracite Region and the Growth of the Runaways

Since the mid–nineteenth century the economy of Pennsylvania's anthracite region, in the northeastern corner of the Commonwealth, relied on extraction of the fuel that powered American industrialization: hard coal. Of the three general categories of coal—anthracite, bituminous, and lignite—anthracite contains the most carbon, burns the most efficiently, and produces the greatest amount of heat. The 500-square-mile anthracite region consists of portions of Dauphin, Carbon, Columbia, Lackawanna, Lebanon, Luzerne, Northumberland, Susquehanna, and Schuylkill counties and contains the largest deposit of anthracite in the Western Hemisphere. Anthracite was discovered late in the eighteenth century and by the mid–nineteenth century played a key role in meeting the country's demand for affordable fuel to heat homes and power factories and railroads. By the end of the nineteenth century, the mining of coal—a large share of it anthracite—was Pennsylvania's leading industry.

Immigrants provided most of the labor in anthracite mines. The United States experienced three major periods of industrial-era European immigration. Between 1845 and 1854, more than 3 million persons, the majority of whom were of Irish and German descent, arrived. From 1865 to 1875, Irish and Germans along with people from England and Scandinavia composed the bulk of the immigrant population. And from 1880 to 1930, approximately 27 million people from Eastern and Southern Europe immigrated. The anthracite region attracted immigrants during all three periods. As the coal industry expanded during the nineteenth and early twentieth centuries, mining became the livelihood of tens of thousands of people from the British Isles, Germany, and Eastern and Southern Europe. Most newcomers earned meager livings in such cities as Wilkes-Barre, Scranton, Pottsville, and Hazleton and in hundreds of small mining towns. Some originally intended to work for a short period and return to their homeland. Others emigrated with more permanent intentions. Though they were now living in the New World, most maintained many Old World customs and traditions. Distinctive ethnic communities formed throughout the region, their cultural life centering on family, work, and religious practices.[24]

Luzerne County's Wyoming Valley, in the northernmost of the four anthracite coal fields with Wilkes-Barre as its major city, emerged as one hub of immigration. Ethnic enclaves of Italians, Poles, Ukrainians, Slovaks, Jews, and others came to join preexisting communities of English, Irish, Welsh, and Germans. The area was originally inhabited by Native Ameri-

Pennsylvania's anthracite region. (From E. Willard Miller, *A Geography of Pennsylvania* [University Park: Penn State Press, 1995].)

Breaker boys at the Kingston Coal Co. colliery playing football, ca. 1910.
(Courtesy of Pennsylvania State Archives.)

cans and, in the latter 1700s, by British settlers from Connecticut in search
of fertile farmland.

At its peak in 1917, the anthracite industry employed over 175,000
workers, who produced a record 100 million tons of hard coal. Yet by the
1920s anthracite's decline had begun. Two decades of internecine conflict
between coal operators and the United Mine Workers of America, competition from natural gas and fuel oil, and inadequate investment in technology were exacting a toll on the industry. By the time the Great Depression
struck, it was clear that anthracite was in serious trouble. After some resur-
gence during World War II, the downward spiral continued into the 1950s,
when fewer than 30 million tons were produced. In 1959 deep mining
ended completely and abruptly in a large portion of the once highly productive northern field around Wilkes-Barre when mines flooded and

Boys employed by the Plymouth Coal Co., ca. 1915. (Courtesy of Pennsylvania State Archives.)

closed as a result of the Knox Mine Disaster.[25] With mining in steep decline, the garment industry filled an economic void and became a key source of employment and family income.

As anthracite's demise appeared imminent, the out-of-town problem became critical in Pennsylvania. According to the ILGWU:

> It was not any longer a question of individual shops escaping from New York, but of a wholesale exodus. "Out-of-Town" has become a big production market, employing thousands of workers. As the demand for low-priced garments increased with the breakdown of purchasing power of the general population during the past four or five years, the small towns in Connecticut, New Jersey, and Penn-

sylvania became practically flooded with contracting shops, all of them working for New York jobbers.[26]

In the 1930s, the union expanded its definition of out-of-town to include a territory covering a 100-plus-mile radius from Manhattan, where it estimated that over 25,000 workers were employed in a virtual sweatshop swamp. The union's Philadelphia Joint Board reported that Philadelphia jobbers had begun to follow the lead of their New York counterparts by sending work to nonunion contractors.[27] A 1933 general dress strike in Manhattan, Philadelphia, and out-of-town areas brought some organizing success, increased wages, and shorter hours. But problems remained as contractors continued to seek lower-cost production. The anthracite region and particularly its northernmost reaches around Wilkes-Barre and Scranton—which contained high population concentrations and afforded relatively easy access to metropolitan markets—were prime locales for contractors who produced dresses and children's clothing for New York jobbers.

As in most regions dependent on a single industry, anthracite's decline and eventual demise left many families struggling to survive. Faced with few alternatives, they turned to public assistance, permanent out-migration, commuting to neighboring states or cities (mainly by fathers and sons), and employment by mothers and daughters in the burgeoning garment business. Although some homeworking took place, most manufacturing was done in contract shops.

Wyoming Valley garment factory owner William Cherkes explained the growth of the area's runaways: "What happened in the thirties and forties, a lot of people [contractors] decided that they could do better outside of New York City by not being controlled [by the union]. So they migrated to Connecticut, Pennsylvania, and Massachusetts to open their plants without interference from the ILGWU in New York City. And by coming here where there was no unions, or no local branches, you could work independently."[28]

The Wyoming Valley was one of many areas in the anthracite region—and in Pennsylvania—that attracted garment contractors, mainly because of its geographically concentrated population, according to Cherkes: "You have to remember that the dress industry went from Wilkes-Barre and Kingston [in the Wyoming Valley] all the way to Harrisburg, and Shamokin, and Pottsville, and Shenandoah, and Hazleton, Scranton, and Carbondale, and up the New York State border and throughout the state. Yet people [factory owners] who came into this area, the Wilkes-Barre area, they would already have a large, available market of labor."[29]

Leo Gutstein, who succeeded his father as owner/manager of Lee Manufacturing in Pittston, confirmed Cherkes's explanation:

> [My father] came here in the early thirties. He came out running away from the union that was organizing in New York City. He had a factory in New York City and he came out here with my aunt, who ran the floor for him at the factory on Twenty-third Street in Manhattan.
>
> He came here, first of all, because there was a large labor force. There was a large labor force of women. The mines were struggling at that time so there was cheap labor out here. People were very reliable, very industrious. In fact, that's how many of the factories evolved in Pittston.[30]

John Justin, an early organizer and educator for the ILGWU in Pennsylvania, concurred: "They moved here because there was enough unemployment, and even though the miners had a reputation for being militant, the mines were closing. Many of the mines had been closed for one reason or the other. And there was no pay. So women went to work in the factories."[31]

Factory owners benefited not only from a large pool of available labor but also from a strong work ethic, according to Cherkes: "They were very industrious, hardworking people, because at that time, to work they would get ahead and do better. And they knew that if they did a better job, you'd respect them and pay them more and do better by them. Because the man, the miner, had no jobs, there were no jobs for men. They were the providers—the women."[32]

According to one local newspaper, conditions in at least some of the region's garment factories were far less than ideal. In 1933 Elizabeth Lynett, daughter of E. J. Lynett, prominent owner of the *Scranton Times*, went undercover to report on working conditions, wages, and treatment of workers in apparel making. After short employment stints at Pell DiMauro Manufacturing Company and Faultless Pants Company in Scranton and Dutchess Underwear Company in Old Forge, Lynett published a series of articles that drew a vivid picture of life in the sweatshop.

Lynett uncovered 55- to 60-hour workweeks; wages that ranged from $2.50 to $11 per week, with the majority of workers at the lower end of the pay scale; routine wage reductions for even the most skilled and tenured employees; unpaid "learning periods"; poor sanitary facilities; routine discipline of workers, including physical abuse for the slightest infraction; and docking of pay for unsatisfactory work.

In her exposés Lynett identified by name well over a dozen sweatshops in the Scranton area alone, carefully distinguishing them from "legitimate" garment employers who treated workers fairly and paid decent wages. She called particular attention to the fact that some of the sweatshops were recruited to the area by the Scranton Chamber of Commerce. Her work drew sharp criticism from factory owners and Chamber leaders and numerous letters of support to the *Times* from disgruntled workers and civic organizations opposed to sweatshops.[33]

Complicating matters further was organized criminals' use of garment manufacturing as a legal front for illicit activities. Criminal elements were known to have infiltrated New York's apparel industry, and by the Great Depression their influence was felt in the Pennsylvania apparel industry, especially in Pittston, a coal town located between Scranton and Wilkes-Barre. According to the Pennsylvania Crime Commission, "In Northeastern Pennsylvania, nonunion garment centers sprung up to take advantage of the unemployed coal mining population used to low wages and poor working conditions. Major Cosa Nostra leaders from Pennsylvania and New York have dominated the industry."[34]

Legitimate factory owners knew about organized crime in both the New York and the anthracite region garment business. Leo Gutstein discussed this influence and the way his father taught him to deal with it:

> [Organized crime] was a fact in the industry. Pittston was an area that was controlled. But Pittston was just an extension of New York

City as far as the garment industry was concerned. The controls came out of New York. The industry was controlled for many years, especially the trucking portion of it. These individuals sort of put a control on the industry. They took a price for that control. They charged a tariff for that control. My father grew up in the Lower East Side of New York. He came out of a very tough neighborhood. He had one piece of advice he had given me as a kid: If you ever get asked for a favor by somebody who is unsavory, to the best of your ability, do it as long as you are not putting yourself in jeopardy or doing something illegal. But never, ever ask for a favor. Once you've asked for a favor, you are obligated![35]

Though organized crime actually owned, controlled, or influenced a minority of factories in the anthracite region, the syndicate would occupy a great deal of the ILGWU's attention. Whether the shops were operated by people with legitimate interests or not, their rapid growth presented the union with a major organizing challenge. According to data from the Pennsylvania Power and Light Company—one of two major suppliers of electric utility service in the Wyoming Valley—between the late 1930s and early 1960s it serviced at least ninety newly established garment factories. And, according to the ILGWU, by 1950 at least forty factories—all contractors—operated in Pittston alone. It is likely that about 10 to 20 percent of the Pittston factories were either directly owned, controlled, or in some way influenced by organized crime.[36]

Equally alarming to the ILGWU was the fact that by 1937 nearly one-third of all cotton dresses sold in the eastern United States were manufactured in contract shops, as were about 50 percent of all children's apparel. A substantial part of this productive capacity could be found in the 15,000 garment workers that the ILGWU estimated toiled in contract factories throughout all of Northeastern Pennsylvania.[37] Adding to the union's concerns was the nature of the production system in these factories—a system that suppressed wages and, as a result, took jobs away from New York.

New York dress factories commonly employed the "whole garment" system of production. That is, a garment worker assembled an entire dress from start to finish. The whole-garment system was used for the making of higher-priced clothing of good quality. In the anthracite region most dress factories employed the "section work" system. The production of a garment was divided into approximately ten specific operations, as on an assembly line. Each worker performed only one task repeatedly—such as sewing pockets or pressing the finished dress—as a garment moved through the production process. The section system was used for the production of large quantities of lower-priced clothing that typically was not of a quality equal to those manufactured under the whole-garment system.

Compensation for workers in whole-garment system shops varied from piece to hourly rates. Workers in section shops usually earned piece rates. The drive for low cost, rapid turnaround, and large output fueled the growth of the section-work system. Piece rate compensation in section shops—often as much as 50 percent less than in New York dress factories[38]—perpetuated rapid turnaround and large output, since the more a worker produced, the greater was his or her compensation. The low-cost competition from the growing number of contractors in the anthracite region had a great impact on the New York dress industry as jobs were siphoned away. The New York Dress Joint Board, the ILGWU affiliate representing various trades associated with dress manufacturing,[39] reported that from 1946 to 1956 its membership declined by over 10,000 as a direct result of jobs going to lower-wage runaway shops in Pennsylvania.[40]

Indeed, runaway factories were of such heightened concern to the ILGWU, Dubinsky told *Time* magazine, that union organizers would pursue contractors who fled to places like the anthracite region in quest of cheap labor as relentlessly as bloodhounds. He boasted that "there are some in hiding, but not for long."[41] By the late 1930s and early 1940s the ILGWU was preparing to respond aggressively.

2

The ILGWU's Response to the Runaways

No other state in the land offers such shocking contrasts as does
Pennsylvania. Here beauty and squalor actually lie side-by-side.
—David Gingold, 1936

The atmosphere in the town was that everything was
controlled and the women had no say at all.
—Min Lurye Matheson, 1988

They were really sweatshops. They stood behind you and
timed you. If they didn't want you, they got rid of you fast.
You couldn't complain about a thing.
—Minnie Caputo, 1993

While it was clear to the ILGWU that New Jersey, Delaware, Rhode Island,
New Hampshire, Vermont, and Maine were all potential battlegrounds in
the struggle to unionize runaway garment factories, it was Pennsylvania,
particularly the northernmost reaches of its anthracite region, that posed
the most significant challenges. David Gingold, leader of the Pennsylvania
organizing campaign, recounted:

Pennsylvania alone with a population of more than eleven million
and an area of 46,000 square miles carried the reputation of "Work-

shop of the World" since its coal, oil, gas, and steel had pioneered many of America's modern industries. The anthracite region, sitting in the northeastern part of the state, climbs the high timbered ridges and spills into the deep valleys. Here were to be seen the exotic onion-domed church steeples of the Slavs. Here were heard the musical accents of the largely second generation Italian, Polish, Welsh, Ukrainian, and Hungarians picking up life where their immigrant parents had left off.

Now the production of women's and children's garments was sweeping in and that variety of ethnic backgrounds was to present puzzling challenges to union organization. This area was now recognized as the area of highest potential in the vast northeast.[1]

In 1935 Dubinsky asked Elias Reisberg, general manager of the Philadelphia Dress Joint Board, and David Gingold, a Manhattan garment worker and former ILGWU vice president, to organize nonunion apparel factories in the northeastern United States and particularly in Pennsylvania. Reisberg was appointed head of the union's Cotton Garment and Miscellaneous Trades Department and Gingold supervisor of the Pennsylvania organizing drive.[2]

Gingold assessed the Pennsylvania situation in *Justice,* the union's official newsletter:

> No other state in the land offers such shocking contrasts as does Pennsylvania. Here beauty and squalor actually lie side by side. Here are filthy mining towns with their warped houses and crooked alley streets where poverty is the byword and death the emancipator; where thousands are slaving in misery while the abundant gifts of nature, denied to those who toil, are generally heaped upon the countryside.
>
> In recent years the chiseling, runaway garment manufacturer has also come to prey upon the poverty-stricken industrial workers in this setting. Pennsylvania is fertile territory for the garment chiseler . . . forever seeking a cheap labor market and to exploit the helpless.

The ILGWU is attempting to establish collective bargaining as a means of raising the standards for the women's garment workers. No spies and private detectives have deterred this union from its forward course in Pennsylvania. There are new forces moving in Pennsylvania and a new interest is manifest in labor organizing. One cannot but feel the stirring of the masses, the slow grumbling of the down-trodden whose sisters and wives have become the garment workers. This is the new spirit in Pennsylvania. This is the forward march of progress in the Kingdom of Coal.[3]

Gingold further recounted his views on Pennsylvania and its anthracite region:

The state of Pennsylvania [was] a world unto itself. Pennsylvania farms are very rich farms. There you had Pennsylvania Dutch. You go a little bit farther and you hit the Mennonite country. Quakers. The mining region [was] a world unto itself. Here you had Poles, Russians, Serbs, [people] from Eastern Europe. The anthracite region was as different from the bituminous region as day from night, [though] the composition of the people, of the workers, are almost the same.

The reason we came to Pennsylvania was because of the fact that, in the thirties, the conditions were very bad. In Pennsylvania the wages were low. Four, five, six, seven dollars a week. The mines were in trouble. In the anthracite region very many of the mines were closed in the early thirties. The only way they did make a living is when two or three of their daughters or wives went to work [in the garment factories].[4]

In Gingold's view, much work needed to be done in the anthracite region, an area he characterized as "a wasteland of the unorganized with thousands resigned to grinding exploitation for the sake of a few slices of bread."[5]

In 1937, at the suggestion of Reisberg and Gingold, the union's General Executive Board chartered Locals 109 and 131, representing dress- and cloakmakers, in Scranton. Also created was the Scranton District Council, with responsibility for the northern anthracite region. Local 249 was chartered in the Wyoming Valley as part of the larger Scranton district along with Local 356 in nearby Old Forge. Additional locals were chartered in the southern anthracite region at Pottsville and Shamokin in Schuylkill County. Because apparel was also a growing industry in other cities and areas in eastern Pennsylvania, locals were established in Allentown, Easton, Harrisburg, Reading, and Stroudsburg.[6]

Since a large share of the anthracite region's apparel manufacturing consisted of women's and children's clothing, the new Scranton district and its Wilkes-Barre and Old Forge subsidiaries came under the jurisdiction of the New York Dress Joint Board. Reisberg and Gingold agreed that their strategy would consist of spreading the ILGWU message as widely as possible, recruiting local supporters who might help in organizing, and conducting selective strikes to draw attention to conditions and increase union membership. The goals were simple: union recognition and contracts with as many runaway employers as possible. As a large percentage of the garment work in the area resulted from contracts with unionized New York jobbers, Gingold concluded that these "jobber barons" were a major cause of the runaway problem. Since many of the jobbers were themselves unionized, there was no reason in the union's view why the runaway contractors should not be organized as well.

One of the union's first major tests in the northern anthracite region came in late 1936, when it struck the dressmaker Pioneer Manufacturing Company in the Wyoming Valley. Workers staged a twenty-one-week walkout, demanding higher wages, shorter hours, and recognition. Some organizers and strikers were arrested and jailed. Gingold recalled this momentous event in the union's Pennsylvania history:

> The shop had over five hundred people. More than half walked out. [It lasted] twenty-one weeks. I used to give them four dollars a week in benefits. Strike benefits. The police became very wild. There seemed to be a desire on the part of these people to be brutal. They came on horseback and they shoved the women pickets off the sidewalk. A couple of women were injured.

We organized a committee of sixteen, a community committee with a rabbi, clergyman, community-minded people. The chairman was the mayor. They went over to him [the shop owner] but couldn't budge him to settle. Then I organized twenty-five miners. They went up to the plant. After that, they couldn't operate much! He came to the committee and we came to a settlement. A very poor settlement: a union, a small increase, a holiday. But our people were in seventh heaven. It was a big victory in Pennsylvania.[7]

After the settlement Dubinsky visited Scranton in February 1937 and headed an organizing rally that drew an estimated 1,500 workers, reportedly "the largest meeting of garment workers yet held in this section of Pennsylvania."[8]

Another important confrontation occurred in 1938 at the Dutchess Underwear Company in Old Forge, a few miles south of Scranton. The company refused to negotiate with the ILGWU or to grant the wage and hour concessions demanded by the union. Gingold encouraged its 500 workers to strike. Nearly all responded, idling the factory for five weeks. The owners subsequently agreed to negotiate with the union. The discussions yielded an agreement for a closed shop and wage increases. The union once again celebrated at a large meeting in Scranton, chaired by Dubinsky, who explained that the ILGWU would not waver in its commitment to organize runaways in the anthracite region.[9]

Despite some successes, by the early 1940s the union leadership recognized that the more populous cities of the northern anthracite region posed a significant challenge as garment factories proliferated. With the passing of Elias Reisberg, David Gingold became head of the Cotton Garment and Miscellaneous Trades Department. The new vice president was eager to focus his energies on the region's runaway shops. Gingold estimated that thirty or more contractors were in business in the Wyoming Valley alone, yet the ILGWU had organized a mere six for a total local membership of about 400 (see Table 1). The union estimated that several of the dozen or so shops in Pittston were controlled by notoriously antiunion and unsavory elements who acted in concert to expand their enterprises by driving down wages and avoiding union interference. Pittston, Gingold argued, provided a moral challenge to unionization in the entire state of Pennsylvania.[10]

In 1944 Dubinsky and Gingold recognized that strong leadership and a major unionization campaign would be needed if the ILGWU were to succeed in the greater Wyoming Valley. They knew, too, that because organized crime was an issue in Pittston, the union could face potentially violent opposition. They needed an organizer who could devote full time to the task, who knew the industry and the problems of its workers, who could speak publicly, command attention, relate to the wives and daughters of coal miners, and was committed to the ILGWU's cause.

Charles Zimmerman, former Communist turned anti-Stalinist moderate and head of New York's Local 22, and William Ross, a Philadelphia organizer, advised Dubinsky and Gingold that Min Lurye Matheson might be interested in the assignment. Coincidentally, both Min and her husband, Bill, had some familiarity with Pennsylvania, as he had been representing the ILGWU in Sayre, a small town in the north-central region of the state, near the New York border.[11] At Dubinsky's request, Min agreed to lead the organizing campaign in the Wyoming Valley. For the near term, Bill would continue to work in Sayre and assist his wife as he could. To keep their family intact, the Mathesons established a residence in Wyoming Valley in early 1944 and Bill commuted to Sayre. The ILGWU's recruitment of Min

Table 1 ILGWU Membership, Local 249, Wyoming Valley, Pa., 1936–1944

Year	Membership
1936	410
1937	241
1938	230
1939	252
1940	218
1941	260
1942	300
1943	311
1944	404

SOURCE: ILGWU, Census Records, Report no. 1, courtesy Lloyd Goldenberg, UNITE! Auditing/Accounting Department, 1998.

Matheson signaled the seriousness of the union's intentions to cope with the valley's runaway shops.[12]

From Chicago to New York to the Wyoming Valley

Minnie Hindy Lurye was born to a Russian Jewish family in Chicago in 1909. As a young girl she developed an understanding of working-class struggles and the causes of organized labor. Her exposure to such issues came from her father, Max, a cigar factory worker and an organizer for the Chicago Cigar Makers Union. Max Lurye was unwaveringly loyal to the labor movement and to socialist principles. His total devotion influenced at least two of his children, Minnie and William, who both became active in the ILGWU. According to Min, as her friends and associates called her, the struggles of working people permeated the Lurye home:

> My father was always active in trade unions and there were always things happening in our house. Immigrants were coming and there were always meetings. We were sort of a hot spot for any ideology or any viewpoint.
> The immigrants would come and sing their songs and some would be revolutionary songs. I'd listen avidly and eat up all this knowledge. Pop would take me to the big Socialist picnics at Riverview Park. I'd sit up on his shoulders and listen to all these fiery speakers. And so you assimilate a lot of that stuff. And it always stayed with me. It stayed with the whole family, except me [and Will] more so. Every Saturday I'd go to the Chicago Federation of Labor meeting. If there was any kind of radical meeting or a trade union meeting, you could always be sure I'd be there.
> And this was my life. You know, listening and taking part in anything I could.[13]

Immersed in Jewish Socialist labor culture, Min learned firsthand about opposition to organized labor. Among her most vivid memories was a raid on the Lurye home by the FBI and Chicago police in 1919:

We lived on [the corner of] Taylor and Seeley Streets in Chicago. One morning I can remember sleeping in a bed with my sister, my older sister, Rose—we were only twelve or thirteen months apart, we were like twins—and the police came up the stairs. We lived three stories up and we had some boarders living with us. Some were relatives. And they arrested every male in the family and tore the house apart. Now my father was the leader of the Cigar Makers Union and they were planning a strike and I guess it was pretty well known. And the cigar industry was a pretty big industry in Chicago at that time and we did have leaflets and literature. But Mom, who was smart and protective of my father, she had all this material on the top shelf in the pantry with jelly jars in front of it. And when the [FBI] Red Squad hit the house, you know, and started tearing it up and looking for proof of the fact that—they were just trying to prove we were Reds—my mother begged them not to bother her pantry because everything else had been ransacked, but leave her pantry alone. And they did, to our surprise.[14]

For Max Lurye labor activism came at a steep price. It eventually destroyed his marriage. His wife, while initially supportive of her husband, later obtained a divorce because she could no longer tolerate his commitment to the cause and the danger that it posed to the family. It also nearly cost him his life when he became a target of Chicago's underworld. The memory of Max Lurye's near death in 1927 at the hands of lieutenants to the Chicago gangster Al Capone stayed with Min throughout her life:

They [the mob] were taking over the unions. And my father had been blacklisted because he was a cigarmaker and there were so many strikes. Anyhow, he was blacklisted and he couldn't get a job, so he went peddling junk. And he had this little truck and there was this junk yard where all these little mostly Jewish peddlers would bring their bits of junk and get whatever moneys they could that day. Why the hell Capone wanted it, God knows. I don't know. So he sent around Lefty Lewis to take over one of their [union] meetings. And my father didn't speak English but he was very good at speaking Yiddish. These guys came to the meeting and they take over the platform and with the butt of a gun they knocked for order. My father was sitting in the front row with my younger brother, Sy. And this gangster makes a speech and he tells them he's going to be there and they're going to do this and so on. In and around the fringes of the crowd are other gunmen. So my father said to him in Yiddish, "Are you finished"? And he just looked at him. And my father marched up to the platform.

The guy didn't know a wild man like him. And my father kicked the gun right off the stand. And he started to talk in Yiddish and they tried to stop him but the people were yelling, "You talk, Max! Talk, Max!" So he talked. He really got the people going. And they retreated. They decided, "What you going to do with these crazy people"? you know. They'll find some other way. And they retreated.

But when my father came to the junk yard the next day, he was standing talking to a man by the name of Braverman who hadn't been active. He was the father of seven children. They went to shoot Dad and they killed Braverman.

Of course, all the other peddlers were scared except my father, who was yelling for vengeance, that he would get them, he would get them. And then a couple of days later he was driving his truck and he stopped for a light on Sacramento and Roosevelt Road in Chicago. As he stopped for the light, so did a streetcar. They came along and peppered my father's truck with machine gun bullets. Some of them hit him and some of them hit the streetcar.

My father had three bullets in the groin area. By the time I got word and got to the hospital, he was still conscious. And he said, "I'm not going to die, I'm not going to die." [15]

Max Lurye survived the attack. The event greatly influenced his daughter, solidified her commitment to the labor movement, and inspired her as an activist: "I was an extremist. I came out of Chicago. The Haymarket case was very fresh in everybody's minds. We hated the courts. We hated the judges. We hated the police. Especially we hated them. We had certain pro-

Min Lurye (right) as a young radical in New York, 1920s. (Courtesy of Kheel
Center Archives, Cornell University.)

cedures for keeping ourselves pure trade unionists. We would not sit down and break bread with an employer. We would not participate with elected officials who were not pro-union."[16]

In 1928, at the age of nineteen, she fell in love with Bill Matheson, a labor movement devotee and intellectual of Scottish origin born in Canada and twelve years her senior. They met at Sunday meetings of the Chicago Federation of Labor. As a sign of their nonconformist views, Min and Bill established a household without being legally married. With Bill's encouragement, in 1930 Min left Chicago to participate in a textile workers' strike in Paterson, New Jersey. She then relocated to New York and associated with many people who had reputations as radicals in the needle trades.

Min became acquainted with a dressmaker, Jennie Silverman, and the two formed a friendship that lasted a lifetime. Jennie recounted Min's association with Jay Lovestone, former head of the Communist Party in the United States, as well as with Charles Zimmerman. Seeing Min's dedication, in 1934 Zimmerman arranged for her to work as a dressmaker in the garment district and enrolled her in the ILGWU. According to Silverman,

Min came to New York and got to know Lovestone and Zimmerman. She was much younger than the rest of us. We all got to know each other. And, initially, she was kind of frisky and I didn't take to her that well.

Then we talked and she told me how they had lived in a furnished room. She said that there was only one bathroom for so many. And since she's decent, she'd go in and scrub the tub real clean and then walk out and give other people the chance to use the room. Others would come and leave the tub dirty. Well, this bugged me. I was sharing an apartment with other friends and I said, "Why don't we get an apartment together?" And we lived together, in one place or another, until, I guess, she moved to Pennsylvania.

So Sasha Zimmerman, who became good friends with her—you had to work in the industry to be a member of the union, to have input there—said to Lovestone, "I need her to join the union." The three of them must have talked it over, and Minnie came and joined the union. They got her a job for sixteen or eighteen dollars a week or something like that.[17]

Israel Breslow, a Ukrainian immigrant who worked in a Montreal garment factory before moving to New York and joining Local 22 in the mid-1930s, observed Min's career: "Zimmerman, when he became manager [of Local 22] in 1932, had two worries: how to enroll women in the union, and minorities. Minnie Lurye played a role. She was chairlady of the local. Minnie Lurye comes from Chicago. Restless. Ambitious. She was to the left of Zimmerman. She was much more sharp politically than he was on many, many issues. You must not forget that [her] parents in Chicago were Communists."[18]

Min's activism and friendship with Zimmerman helped her to gain the respect of co-workers and union leaders. Within a year she was appointed to Local 22's executive board. In 1935 she was elected "chairlady" of the local and its 32,000 members, a prestigious and important position. A short time later Bill accepted a position as an organizer for the ILGWU. By the late 1930s, they had both embarked on union careers that would span over three decades.

In 1941, when Bill accepted a union position in Sayre, Min continued to work in New York until their decision to have children brought some major changes. They married and their first daughter, Marianne, was born in August 1941. Min and the baby joined her husband in Sayre, and they established a home there. Min fully removed herself from ILGWU matters and concentrated on raising a family. A second child, Betty, was soon born to the Mathesons.

However, Min and Bill never quite felt comfortable in rural Sayre. Not only were they removed from many of their ILGWU associates and the pace of life in New York, but labor organizers and Jews with ties to unions apparently caused some concerns in the town. Min's stay in Sayre ended in 1944 when Dubinsky asked her to "clean up the mess" in the Wyoming Valley.[19] Min wasn't quite convinced that this would be a long-term assignment. According to Betty Matheson Greenberg,

The reason that my parents were in that area [Sayre] was that my father was organizing for the ILGWU. At that time my mother was a housewife. She had left her job and decided she wanted to have kids. And Sayre is quite an interesting place. People came out to watch my mother go down the street to see if she had horns and a tail because my mother was Jewish. They thought that Jewish people had horns and tails! I don't know what really led her to go back [to the union and organizing]. She got an offer. David Dubinsky called her to New York and she said [to Bill], "Let me just talk to him." And he asked her if she wanted to go back to work organizing.

She and my father talked it over and they decided she would. It was at first going to be a temporary thing. She was simply going to work out of Wilkes-Barre to see if she could get anyone to join the union. Well, that was the beginning and the end of it. They hired a housekeeper for my sister and me. My mother at that time paid the housekeeper the same amount that she earned, which was thirty-five dollars a week. And she went back to work because she was a good organizer.[20]

"Everything Was Controlled"

When she arrived in the Wyoming Valley, Min Matheson found working conditions worse then she had imagined. She was struck by the subordinate position of women garment workers, most notably in Pittston:

The atmosphere in the town was that everything was controlled and the women had no say at all. They did the sewing and the cooking and took care of the lunches and got the children out to school and the husbands out to work in the mines. This was their life.

They [employers] told the women, "We'll teach you to sew." They worked for weeks for nothing. And the hours! You know there were laws in the land but they weren't carrying out any of the laws. They did what they wished and made it easy for the women to come in any time of the day or night. Double, triple shifts.

I'd talk to the women at meetings. And the first thing is, "Are you registered to vote?" Yes, they're registered to vote, but they don't vote. "Why don't you vote? Do you go down and vote?" "Well, we do, we do, we go down and we register but we can't cast our vote. Our man has to cast our vote for us." I said, "Why?" "Well, that's the system."

That's the system which the Mafia had ordained to control the elections. The women would go in and sign as citizens, but then the man [husband or another male, possibly a shop owner] would go to the polling place and cast their vote. The women were never allowed to vote. Attorneys and judges, a lot of them knew, but they . . . It was all covered up, you know. They could have stopped it.[21]

Though insulated from the situation to a certain extent by her youth, Betty Greenberg learned about and supported her mother's discoveries:

The girls in the factories not only did their factory work, most of them, all of them, did their housework. And they took care of their kids and made the meals. Their husbands didn't do anything to help. Most of them were out of work. They went and stood at bars all day. That's why the factories grew here, because the men were out of work in the mines. There was no money. And if the women didn't get to work, there would be no paycheck.

The men would stand outside of the factory and take the girls' paycheck when they came out of the factory on Fridays. And a lot of the girls would be begging . . . . A lot of the men drank the money up. "Just let me have enough for the rent, just let me have enough for the rent." When the girls voted, the men went and they voted for the girls.

I mean it was a very, very chauvinistic society. And very accepted. Most of them weren't educated past the third grade.[22]

Min confirmed the ILGWU's assertion that a prime incentive for the emergence of runaway garment factories in the Wyoming Valley was the lack of a strong union presence and the availability of cheaper labor. Organized crime, moreover, was part of the local scene.

At that time there had been things happening with organized crime. The big shots in New York, the Genoveses and Albert Anastasia, were having legal problems. So they wanted a legal front for their illicit operations that included everything. They had really set up a center in Pittston that had the most beautiful girls, you know, little dark-haired Italian girls. Little beauties. Prostitution was rampant. So now they needed a legal front and the dress industry was easy. You need very little capital and all you have to do is have a handful of machines and you're in business. And all these manufacturers in New York who were looking for cheap labor loved it. All the mines were down. Men weren't working. We had organized in New York and surrounding areas. The wages were getting higher, you know. Piece rates were better and employers were looking for low wages and areas where they could produce the garments at the lowest level possible. So they were running. They ran to the coalfields of Pennsylvania.[23]

From the perspective of an ordinary garment worker, the situation in the Wyoming Valley was difficult, but people had few choices in the sagging anthracite economy. According to Minnie Caputo, a garment worker in Pittston,

I started when I was sixteen or seventeen. I worked in a factory over in West Pittston. That was one of the first dress factories around here. It was nonunion. I think I got ten cents an hour. Well, all I worked [in that factory] was twenty-four hours because I think I got two-forty in pay at that time.

He timed you. They were like sweatshops. They stood behind you with the stopwatch and timed you and if one girl did eight operations he would say, "Why not you"? And, you know, not every girl had the same speed. But they timed us for every operation and they compared. If they didn't want you, they got rid of you fast. You were out. We worked hard.

They were really sweatshops. You couldn't complain. You couldn't complain about a thing. You had to do what they told you to do. You had to work. There were no piece rates. If you worked overtime, you never got time and a half. We got piece rates after the union got in here. But the dress industry is what kept this town going. You know, there was nothing here for the men. There was just the mines, but when that went, it was just the women working.[24]

For many women like Minnie Caputo, a "learning period" was a customary part of their initiation into garment factory life. An investigation by the *Christian Science Monitor* in 1933 reported on such tactics in contract shops:

> In their desperate need for work and wages, girls and women jump at the invitation to "sit right down at a machine and go to work"; they do not venture to inquire about wages and, indeed, some are told they must work a week, or even two, as learners. Even experienced workers are having to accept this invitation. At the end of the "learning period" they are told their work won't do and they are then dismissed for a new group. By this trick a contractor can get several thousand dresses for almost nothing.[25]

The situation in which workers found themselves ran contrary to the workplace philosophy of Min Matheson:

> I'm a great believer in American democracy. Having the right to vote doesn't make it democratic. You've got to have a say in your working conditions because that's were you spend most of your life. And if you can't determine your own life in the sense that you have some say in what kind of conditions you work under, you don't have democracy. You are really being denied your democratic rights.
>   I went on the theory that if you let one factory stay nonunion it would soon contaminate the others. The way this town [Pittston] was controlled, we couldn't take that chance. The union was there. They had to be union. We didn't have the people [initially], that's

true. We did not have the people. They [owners, organized crime] controlled them. But we wouldn't give up. I never wanted anything [for myself] and that's the truth. People would say, "What's in it for you, Min?" And I'd look at them because it was so abhorrent to me, the way I had been brought up, to think that anyone would use any situation like this for themselves! We had two children to bring up and we got very low pay from our union and we worked very hard and we weren't in it to get a thing for ourselves and we never did. And we never would. We fought hard to get the people, to organize them and to do the things that would call attention to our plight.[26]

To Min there were clear reasons why workers would support the ILGWU. It was a vehicle to challenge the status quo and to improve wages and working conditions:

There was absolutely no control of hours, wages, and earnings. But the women were smart enough to know that they put in a very full day of work for which they were getting a pittance. So they were very much in our corner since we fought for them. You know, they were afraid to talk up because of all these problems in the background and they were afraid for their men because the Mafia wasn't exactly nonviolent. When I walked in they knew something was gonna . . . Drama . . . drama had arrived. They loved the drama. It was really a question of showing these people that they couldn't run roughshod over these women. So we had our work cut out for us, talking to all of them. And they began to admire what we were doing and agreed with what we were doing. It's like the Messiah had arrived. We were going to change all this. We were like the yeast in the dough. Now we came along and we gave them status.[27]

In Min's observation, women suffered from low status not only in garment factories but in the community generally. Complicating matters was the need for many women to be responsible for holding their households together in difficult economic times:

Actually, you think, this was the forties and women weren't very prominent, weren't expected to know anything except house and children. As we got to know different families, particularly the men couldn't understand how I could know these things or speak on these subjects. The backwardness toward women was unbelievable. Eventually the women were really coming into their own, so to speak. I'd say they were excellent and very capable. Of course, I'm prejudiced. In my eyes most of them were far and above the men they were married to. They had a habit of referring to the good ones. "He's a good one, Min." Some miner was killed in the mines and they'd say, "It's too bad, because he was one of the good ones, Min." The bad ones were the men who drank up their wages, beat the women, who dominated and had tempers like fiends. The women worked and carried on and kept the house together. It was [often] a poor kind of marriage relationship.[28]

Min's views on organizing and improving the lives of workers were not uncommon among people who worked for the ILGWU. The organizer Paul Jacobs has written that "to be a union organizer was one dream of many young liberals and radicals during the thirties and early forties; to be an ILGWU organizer was almost more than one could have hoped for even in their wildest fantasies. And, for me, organizing shops who had fled from New York to escape the union was even more than a mission. The status that went with being an ILGWU organizer meant a lot to me at a time in my life when I was bedraggled and miserable."[29]

Drawing upon her skills as an effective organizer and speaker and her husband's ability as an intellectual and educator, Min worked to garner the resources necessary to establish the ILGWU as a political, economic, and social force. During the latter half of the twentieth century, the Wyoming Valley District of the ILGWU had become the anthracite region's most influential labor organization as well as one of the more significant in Pennsylvania and the vast ILGWU network. By the time Min and Bill departed in 1963, the district consisted of nearly 10,000 members in 168 unionized factories. Moreover, the ILGWU enjoyed the respect of its members, politicians

Min Lurye Matheson, 1963.
(Courtesy of Stephen N.
Lukasik, Lukasik Studio.)

and public officials, business leaders, educators, the media, community activists, and employers, as Leo Gutstein explained:

> The ILGWU was made up of people who were idealists—who came into the movement because it was what they believed in. When it was idealists that I dealt with, I grew to respect them. I understood who they were, who they represented, and what their job was and would deal with them on that basis.
>
> Min was an idealist. Min had a very strong personality. I knew her as a kid growing up. She became a friend of my father. As district manager here, she wielded a lot of power and had a lot of members and influence. This was an area that she had organized and had a lot of control over. She did a good job in convincing the public and many others that the sweatshops were terrible places to

Wilfred (Bill) Matheson, 1963.
(Courtesy of Stephen N.
Lukasik, Lukasik Studio.)

work. The stigma stayed with the industry. And Min organized against that stigma.[30]

As Min Matheson set out to transform the ILGWU into a powerful and widely respected organization, establishing a presence, organizing, and relaying the union's message were among her first significant challenges.

3
Strategizing and Organizing

We were scared. I don't want you to think that we were so brave. We were scared.

—Min Matheson, 1982

I remember hearing her [Min] in Pittston. I was new to all of this at the time.
But I remember thinking, "My God, this woman has a lot of guts!"

—Sol Hoffman, 1998

As she set out to organize runaway garment factories in the Wyoming Valley, Min Matheson realized that she faced numerous challenges. She had to learn about the area's culture and people and secure resources that could aid the ILGWU's cause. The union had to establish a presence. Organizers had to test various strategies in an uncertain environment. Workers had to be educated about the benefits of unionization—a long-term challenge. Fear and intimidation were constant problems. According to Min, "It wasn't that organized crime was exactly nonviolent. We were scared. I don't want you to think that we were so brave. We were scared." The ILGWU did not set out deliberately to challenge criminal elements, but Min soon realized that organizing in Pittston could not be accomplished without some clashes: "We didn't know that they were mobsters. We didn't know who they were. People said, 'How could you take such chances?' But we didn't know until we started to deal with them that they had such reputations."[1]

Min's close friend Jennie Silverman explained: "The garment industry has always had a problem with the underworld. A lot of them were exported to Wilkes-Barre and Pittston, a stinkin' town, if you don't mind. So Min was the manager. And she had to make arrangements with them for the workers. But what choice did she have? They were the employers. Either make arrangements with them or don't make any arrangements."[2]

Among the first underworld figures known to exist in the United States were the heads of the Volpe, La Torre, and Sciandra families, who immigrated to Pittston from Montedoro, Sicily, in the 1880s. Several males in each family worked as coal miners, then contracted and leased anthracite mining operations from larger coal companies and invested the profits in various business ventures. Their businesses often served as legitimate fronts for illicit activities, such as bootlegging alcohol during Prohibition, gambling, and loan sharking. By the 1930s, organized crime invested in garment manufacturing as a legitimate enterprise.

John Sciandra, the alleged syndicate leader in northeastern Pennsylvania during the 1930s and 1940s, and his son, Angelo Joseph, were among the first garment contractors in Pittston. Sciandra's successor was Joseph Barbara, host of the ill-fated 1957 organized crime "convention" in the upstate New York town of Apalachin, which made national headlines when it was raided by law enforcement. Barbara and his lieutenant, Rosario A. (Russell) Bufalino, also owned interests in Pittston garment factories. By the late 1950s Bufalino had become the head of organized crime in the area. He boasted a national reputation as an underworld figure and was dubbed "one of the most ruthless and powerful leaders of the Mafia in the United States."[3] Though the ILGWU had gained a firm foothold by the time of Bufalino's tenure, he would pose problems in Pittston (discussed in a later chapter).

As the ILGWU began organizing, it became apparent that Sciandra and other Pittston elements were adamantly antiunion and could rely on local law enforcement to keep organizers at bay. According to Min,

At the Pittston end of the valley it was as if every empty store, every empty space was occupied by dress shops. So we started organizing. We'd go into Pittston and the police would put us back on the train and tell us, "You are not wanted here." So we started coming in by car because they [the police] used to watch that little railway station to see if we were coming. They [organized crime] could [often] pick the people who would be mayor. They could actually control everything that went on in the town.[4]

Pittston garment workers knew of the presence and influence of individuals with questionable connections and backgrounds. Yet many were resigned to working for their companies. As a young man in Pittston about the time Min began organizing, Anthony "Tony" D'Angelo worked in a factory owned by the Sciandra family. He experienced at firsthand the corrupt relationship between the underworld and local law enforcement:

I was a presser and I was going to barber school at night. So he [Angelo Sciandra] comes around every day and says, "Work tonight, work tonight." I said, "Hey, I can't." Well, he said, "You're fired." So I didn't show up for work the next day and the foreman calls me at home. "Where the f___ are you"? I said, "I'm home, where am I? He fired me." He said, "Ah, he don't know what the f___ he's doing. Come in to work." So I come in to work.

Then he [Sciandra] tells us, "Now you're going to work next Saturday." OK? And myself and Charlie Policare, Carmen Sciandra, and Rocky Scholanchi, we took every dress out of the factory that [Saturday] night. We were making dresses for Sears and Roebuck. We took every dress out of that factory and left the belts on the hangers. Every dress we took out of that factory and took them up the Yatesville [near Pittston] police station. Are you listening to me? *The Yatesville police station, in boxes!* And we stored them on the second floor of the police station up in Yatesville. And the belts from the dresses were left hung on the hangers and eight buttons were dropped at every belt. OK?

So Lenny Blandina [another alleged member of the Sciandra crime family] took every sewing machine head off most of the expensive machines in the factory. And Rocky, Carmen, myself, and Lenny took burnt machines, *burnt machines,* and put them in their place. Machines that were already in a fire. These machines were charred, they were burnt. You could see they were in a fire. We put every one of those machines in place of the new machines and took the new machines up to the second floor of the Yatesville police station!

And three days later, the factory was torched! We didn't know what was happening. We were naive, OK? Because we were young. We were young. He torched the factory and he got paid in insurance for the dresses and for the machines that were lost in the fire. And all the while, they were up in the police station![5]

The presence of gangsters led Min and others to conclude that the area would be very difficult to unionize. In comparing organizing in the Wyoming Valley with his work in other areas of the northeast, John Justin reflected, "It was more difficult to organize here. There was a different element here. The area was more aggressive. There was more resistance by employers."[6]

According to Betty Greenberg, the early years were difficult for Min, yet, with Bill's support, she worked to understand the local situation and to convince people that the ILGWU provided an alternative:

My mother and father were quite good at their jobs because they believed in it. They believed very much in the fact that human beings should not allow themselves to be put in a position that they have nothing to say about what they earn and nothing to say about how they work. We should not allow sweatshops and children working. That was all very much a part of their thoughts. She was trying to do something for people, and really, I think that's what made people accept and believe in her. She never spoke down to people. She never walked away from people. Another reason that they liked her was that she was one hundred percent with them. Never walked away from anybody.

She'd start by getting a list of factories. She might go to ten factories a week to see if you can't break the ice. And then if you can get the girls at all interested, then you might come back and have a meeting. And most of them were so beaten, you know, by the attitudes before they [organizers] even came in.

If [she] was having a meeting and wanted people to come, she didn't say, "We're having a meeting, we hope you come." You've got to do a little work. You've got to tell people why you want them to come and why it's important for them to come. She'd say to them, "I'm one of you. And let me tell you, if you don't join the union you are always going to be where you are now."[7]

Min's leadership was important in winning over workers but she also relied on the anthracite region's larger labor heritage. The United Mine Workers of America (UMWA) had been integral to the cultural and social fabric of the area since the turn of the century. Well before the ILGWU arrived, the majority of anthracite mineworkers were organized. Some miners had even formed an opposition union in the 1930s, the United Anthracite Miners of Pennsylvania, as a result of dissatisfaction with the UMWA.[8] The ILGWU relied on this culturally based support for organized labor, said Min:

The other unions were very happy about the fact that we were organizing and we were making so much noise because the labor movement in this area was always UMWA. The other unions meant nothing. We were always bringing them into the picture just as we brought our members in. I never did anything without asking people from other unions, "Would you come and help us out? Would you do this? Would you do that?" When we ran a meeting I'd have them come down and be a part of it and speak. If we had a little lunch, or it would be Saturday, or beginning-of-the-year meetings, [we would welcome] as many as wanted to come.[9]

Interunion solidarity also came from the International Brotherhood of Teamsters, who hauled raw materials to garment factories and finished products back to jobbers. According to John Justin,

In setting out to organize the shops we used the pressure of getting as many people in the union as possible, maybe the whole shop, so the employer had no choice but to sign. We used the method of getting the Teamsters, who were growing at this time too, not to deliver [raw materials and finished products] and paralyze the employer. If we were under pressure, we would use everything possible, not anything violent, but everything possible to bring it to conclusion.[10]

Tapping into the culture of unionism also meant recruiting unionists who both knew of and were known among local people. To counter the toughness of organized crime, Min secured determined people familiar with the local terrain. One such individual was Angelo "Rusty" De-Pasquale, a member of the UMWA:

Even before I met Minnie I was with the UMWA. Yeah, Christ, I was always union. Since I was a kid I worked in the labor union. Like I say, Minnie asked me to go [and help the ILGWU] on account of me being from Pittston.

It must have been at a union dance or something, because I used to go to all these dances. One of the girls introduced me to Minnie and she asked me about Pittston. She knew that I knew my way around and all that. [Later] I met her and we went to Pittston and she showed me the shops and everything else. I said, "Look, Minnie, you want me to go along, I'll help you. Whatever I could do for you. I don't know what the hell you're talking about, but I'll help you." What the hell did I know about organizing, you know? Then she sent me out organizing some of the girls that I knew, to try to get them signed up.

And then, once you get going, then it comes to you. I'd go to people's houses and walk right in. And they say, "Who the hell are you?" and stuff like that. And I tell them who I was. I'd get them signed up. A lot of them would say, "Go to hell," and I'd go to hell. What was I going to do?

Oh, the way she talked. The way she spoke to people, the way she treated me. You know, she'd do anything for me. She was a good woman. Minnie would come up to the house. She'd have dinner with me and all. Yeah, we were close.[11]

Inevitably the ILGWU had to deal with legal problems. Daniel J. Flood, a former stage actor turned labor-friendly attorney, provided legal support to the union. Rusty DePasquale explained that after one particular Pittston organizing drive in the mid-1940s,

> [the police] locked me up. They brought me in for a little while. So Minnie come to me. She said she was new here, she didn't even know about lawyers or anything like that. She said to me, "Rusty, where could I get a good lawyer?" And I gave her the name of Dan Flood. He was a lawyer at that time and he was a good lawyer. One of the best. So she got him and I will never forget.
>
> I'm locked up downstairs and Dan Flood come down with the chief of police. Ten minutes. He wasn't there ten minutes. Dan Flood asked me, "What's your name?" and all of that and I told him. And he said, "What are you doing here?" And I said, "You ask him [the police chief] what I'm doing here!" My ear was still bleeding. When one of the cops hit me, he cut my ear. I said, "You see, they hit me over the head."
>
> So he said to the cop, "What's he doing here"? He said, "Well, I got him charged for starting a riot." And he said, "Well, who do you have that he rioted with?" And with that, the cop got all messed up. The chief released me right then and there.[12]

From the early days of establishing a foothold to later years when the ILGWU matured, it sustained a strong relationship with Dan Flood, particularly when he was a powerful member of the U.S. Congress (discussed further in Chapter 5).

As she built resources, Min expanded the effort in Pittston at a factory affiliated with unsavory elements:

> We picked a factory on Main Street, 77 South Main Street. The owner's name was Falzone. And the irony is that some years later we [the ILGWU] bought that building, that very building.

You can imagine the sensation it created when I arrived with a bunch of women and we were going to organize the factory. Of course, the women who worked in the factory were scared out of their wits. But we had leaflets and we had a message and we stuck by it. And then he'd call the police. The police were very mean and nasty. One of the police officers grabbed me and put my arm behind my back. And I said, "What are you trying to do, break my arm?" He wouldn't let go of my arm. It was very painful.[13]

Rusty DePasquale explained that organizers learned two lessons rather quickly: physical violence would be used against them and they needed to test appropriate responses:

See, I was sort of a nutty bachelor. I wouldn't take no shit from nobody, you know. Anyhow, the girls are walking up and down with the signs. And there were a couple of wise guys [members of organized crime]. And they came over and they razz the girls and all that shit. And I said, "Now look, let the girls alone because you are going to get it one of these days." And the cops come over. I had a guy organizing with me by the name of Johnny Justin. He was from New York. He was a Jewish kid, Jewish folk.

Anyhow, the cops come over and they grabbed Minnie and they grabbed this guy [Justin] and threw them in the [police] car. When I saw that, the way they mistreated Minnie, I thought, "You son-of-a-bitch." I ran over to the car and the one cop, who was Irish, I grabbed him and I pulled him out of the car. I even cursed at him and said, "Son-of-a-bitch, you don't do that to a woman. I'll punch you right in the mouth." You know what I mean? And then the other cop got out with his club. And then some of the girls got in between. Anyhow, they took Minnie and John Justin for a hearing, or whatever the hell happened.[14]

As Pittson's organizing efforts continued, blatant threats came from the town's criminal forces. According to Min, on one occasion when a carload of organizers and picketers arrived in Pittston, "there's [Angelo] Sciandra

Min Matheson on a picket line with Local 249, Pittston, late 1940s. (Courtesy of Alice Reca.)

sitting on top of his car and he's blowing Taps for us [on a trumpet]. He was a musician. He has a big overgrown kid, you know. He was very young in those years. But it wasn't a very happy thing because, first of all, the war [World War II] was on and, you know, Taps. We thought, 'They're going to kill us today, listen to this.' So Sciandra was sitting up there on the car blowing Taps for us. We would fight and catcall and tell them off."[15]

Dorothy Ney, another Matheson recruit, recalled that mobsters would "watch us, scare us. They threatened us. If we passed them by, they'd make a remark. They were doing what they were told to do. I remember one fella [an organizer] who was beat up. But why he got beat up that day I have no clue. That poor guy came out on the picket line and they really beat him up. He was bleeding all over the place." [16]

One strategy was to win over factories not influenced by organized crime. In 1945 the ILGWU's Northeast Department (renamed from the Cotton Garment and Miscellaneous Trades Department) decided to launch an all-out effort to organize McKitterick-Williams, a contractor that operated in four states with over 800 employees. One of its facilities was located in West Pittston. David Gingold authorized "simultaneous stoppages in all of these shops" and directed Min to organize the factory:

> See, interestingly, there were legitimate employers. There was a big employer from New York called McKitterick-Williams. They had a shop in West Pittston but they were operating nonunion. While we were trying to organize [mob-controlled shops] we were also trying to organize them. We always bit off a little too much!
>
> There was a manager running the place. He was a little sympathetic to us. He was not against us and they really ran a nice factory. Jimmy Boyle, that was his name. And Jimmy didn't care if we came in. He used to come out and talk to us.
>
> We had pickets in order to find a way to get the union in. The women were very friendly and sympathetic to us. We had some of the shop out on strike. Some of the really good operators joined the picket line. And we were working on this and on Main Street in Pittston which was really my aim, because Main Street in Pittston was controlled by the tough guys. They were the shops that the tough guys controlled. We knew the tough guys controlled them because the women, you know, they knew. [17]

Efforts at McKitterick-Williams were successful and added several hundred garment workers to the ILGWU's ranks.

Min employed another strategy in building the union: altering the status in which women garment workers were held in the community. In one instance, she led an effort to challenge the practice of deferring women's voting rights to men. Organizers relied on rank-and-file members as well as advocates from other areas of the valley who traveled to Pittston to picket for the cause. The union educated local members regarding the franchise, then encouraged them to vote with ILGWU backing at the polling place. This was one of the first times Min used the local radio to the union's advantage:

> The women were never allowed to vote [in Pittston]. Well, I said, "Will you explain that to me? Do they send the ballot to the house? What do they do?" It seems that the mob controlled the polling places and the man [husband or owner] would go in with the woman and they would sign in like they should and then the man would go in and cast the vote for both of them! So it was "My man votes for me."
>
> Here I am telling them, giving them our propaganda, and they vote for whoever the mob was supporting. The first time I heard that I said, "Well, we'll just see about it." I would tell Bill, and he was so good at writing these little speeches with a little humor.
>
> So I took time on the radio. I'd start out by saying, "So you think it's a free country? So you think you have the right to do this or that? But you don't know anything if you don't know the town of Pittston. They have their own laws down there. We're going to put Pittston back in the U.S.A. I'm serving notice on those who think they control Pittston politically, that we're going to put Pittston back in the U.S.A. What do I mean? No man is going to vote for a woman. I have a personal pledge of women that they will go to the polls and they will sign in but they will also cast their own ballots!"
>
> So in that election we picketed that polling place, which was the hotbed for tough guys. Oh, they were murder against us because this would take their political power away. And we took women from down here [other areas of the valley] because the Pittston women were afraid. One woman [at a time] would go into the polling place and when they told her she couldn't vote—her husband would vote for her—she would refuse. She wouldn't sign the roster unless they agreed that she would vote.

That was one of our first big fights. That was a bigger revolution, I think, than organizing the shops.[18]

A third tactic was to position the ILGWU as a legitimate member of the larger community and identify the union as a supporter of the common good. Part of the strategy meant working with prominent community leaders. Min thought that such efforts might aid the union in its struggle, yet it required her to reexamine her own biased views toward capitalists and business owners as "the enemy." Although Min gradually abandoned the virulent anticapitalism of the ILGWU's far left—because she grew to believe that the American economy was capable of progressive reform—the idea of cultivating the support of community elites did not come easily:

> So here we had to begin to learn that in order to organize, I've got to have the community in my corner. Gradually we very carefully participated in one community venture or another. I remember the first time I had to sit down at a meeting in the Sterling Hotel. One of our employers, who was heading the Employers' Association, asked me if I would come to this luncheon. I was very hesitant because I had been brought up strictly where labor doesn't mix with capital. You don't betray your workers and sit with them and eat with them. I had all these very different concepts.
>
> So I went with great trepidation to the meeting. And I met Jack Sword, who was then head of the Chamber of Commerce, and this head of the Employers' Association, and Frank Anderson of the Miners' Bank. My first reaction was, "What am I doing here? What if somebody gets a picture of me and sends it to New York? They'll throw me out of the union!" You know, I had been brought up with such a rigid framework. But it was an interesting conversation. What they were trying to do was to get our organized shops to participate in the Community Chest. That's what the whole meeting was about. It was some time after the war. That did us a lot of good, which I didn't realize at the moment. Because once we got to that phase, we were sort of accepted by the elite. Not that I wasn't walking on eggshells at first.

We worked with them. We later formed the Committee of 100 [for economic development] and I was on the committee. I think I got more of an education than they did! I told you I had a very lop-sided background in training that was all labor. I didn't know from the other side. I didn't know about business problems, small business, big business, industrial troubles. I only knew labor's end. And it is important to have a more rounded view of the problems. We [became] active in the United Fund. We were active in the Red Cross, donating blood. And they admired what we were doing. We took leadership and told our people, "This we should do." I think they [community elite] supported us because they thought it was good for the community to have an organization that was community conscious and upped the earnings of people, which made it better for the community. They were for us.[19]

The decision to participate in community projects signaled an important evolution in Min's organizing methods and policy. Confrontation had been the chief tactic but gradually she realized that cooperation could be equally beneficial. Local business and community leaders saw the economic benefit of obtaining the participation of ILGers in, for example, the Community Chest, which raised charitable contributions for human services. Likewise, Min began to see community participation as an avenue through which the union could enhance its legitimacy. In time the union's involvement in numerous aspects of community life became as common as its organizing drives and strikes.

While Min was undertaking these strategies, the union's general executive board made it clear that Pittston remained a key battleground:

The drives launched by the Northeast Department were aimed chiefly at the new nonunion firms which had come into the industry taking advantage of cheap labor. This tendency was typified by a group of garment employers who set up plants in Pittston, Pa., planning to operate them with low-wage labor and to mobilize some notoriously anti-union elements in their support. The Pittston conflict [is] a challenge to the entire union structure in Pennsylvania. Union organizers faced severe counter-fire from nonunion employers which did not hesitate to use strong-arm tactics in order to intimidate the organizers and workers.[20]

Reality called for a fourth strategy: persistence in the face of virulent opposition. Though Min never condoned physical violence, she had to remain creative and strong in the face of threats. On one occasion organizers demonstrated that they simply were not going to remain passive in Pittston. According to Min,

> We had to evolve and think up ideas how to protect ourselves. Another girl we got from one of the shops, Helen Barnosky, I think they're Lithuanian . . . anyhow, she was driving the car and we were going down to a very early-morning picket line and there on the corner were all these tough guys. She said, "I don't think we better stop because we are outnumbered." What were we, five women in the car? I said, "Helen, no matter what happens today, we have to stop. Because if we don't turn up today, this may hurt our chances to really get this town organized. We have to just put up a very brave front." I said, "But we have to think of something spectacular to worry them like they think they are worrying us."
>
> But we couldn't figure and we didn't have much time to talk because they saw our car coming down. Then she turned around and we were coming up their side of the street in order to stop in front of the factory. It's as if we all had one mind, that if they were going to get tough and beat us up or do something rotten, really rotten, then we wanted people to know about it. So the minute we stopped, as we got out of the car we were all screaming. I got out of the car and I said, "You rotten hoodlums! What are you doing in this town? You don't live here. We live here. This is our town, not yours! And you do one little thing to hurt these women . . ." because I was their leader.
>
> In the meantime all the other girls were screaming at them. Screaming at them! Pretty soon windows were opening and people

were putting their heads out. I said, "There are witnesses to anything you think you are going to do." And honestly, the men almost
went crazy. It was like, "My God, how can you do anything with a
bunch of crazy women like that?" They were walking around, wav-
ing their hands, putting their hands over their ears. Not a squeak
out of them. Nothing. You see? So I always said that the women
defeated them.[21]

On another occasion, as hostile words were exchanged between toughs
and pickets lined up across the street from a Pittston factory, one alleged
mobster, in a fit of anger, shouted that Min should bring her "weakling husband" to the picket line and see how long he'd last. In a forceful show of
resolve, Min steamed across the street, singled out the mob boss Russell
Bufalino, stuck her finger in his face, and shouted for all to hear, "I don't
need to bring Bill up here, Russ, because I'm twice the man you'll ever be!"[22]
On another organizing drive, gangsters verbally attacked her as a "slut"
unfit to lead the union. To demonstrate her resolve once again, Min furi-
ously phoned the union hall and asked a friend to collect her preschool
daughters, dress them in brightly starched pinafores, and deliver them to
the picket line. Once they were there, she handed them picket signs and
ushered them to the line. How could the tough guys call her a slut now?
This event stands out among Betty Greenberg's childhood memories:
"Then there were the picket lines. There were some serious times, that it
was important for my mother to show that she wasn't afraid and that she
had a lot of belief in what the girls were doing. She took my sister and I to
the picket line in Pittston. My father was going to shoot her! He was furi-
ous with her. And later she realized he was right."[23]
Rusty DePasquale remained equally uncompromising when the need
arose, though he would not dismiss the use of violence if, in his view, it
was needed. During the 1958 general dress strike he attempted to stop a
mob-owned truck from delivering finished goods to New York:

It was tough in them days. Christ, yeah. Especially when you were
muckin' around with that Pittston gang. One night I got called up.

Well, it was about three o'clock in the morning when we had a strike going on down there. I had two guys watching the shop, you know. They call me up and said, "Come down here. There is a truck here loading stuff." I said, "Well, stop them." They said they were afraid. So I went to the strike headquarters and I asked Bill Page to describe the truck. I said, "Get into the car."

So anyhow, . . . when I'm going down I see a truck about three-thirty or four o'clock in the morning. I followed the truck, and sure enough! We went up [route] 115 toward the Effort Mountain [a road to New York]. It was snowing and all. I wanted to get the truck and unload it, burn it, or distribute the clothes to the poor people. I was going to do that but he [the truck driver] kept pushing me over [off the road]. I never knew that it was a Mafia truck.

I happened to have a gun in the car. So I thought, "You son-of-a-bitch." I wanted to give the gun to Bill Page but he got under the dash. He was a big sucker. He crawled under the dash! I had a Packard, you know. "Please, Rusty," he kept saying. But I said, "Get down." He said, "No, no, no!" So I started shooting over his shoulders and driving. The guy was pushing me over and it was snowing and the road was only a two-lane highway then. That's all it was, a two-lane going up 115 up toward Effort. He pushed me over, you know! But I had this son-of-a-bitch of a gun.

I couldn't do anything [to stop him] so I turned around and I come back. I didn't know it was a Mafia truck.[24]

DePasquale occasionally recruited the support of acquaintances who could be as tough as any group they were opposing, though Min did not necessarily approve of his tactics:

She'd even give me hell. One time I said, "All right, I'll burn that son-of-a-bitch. I'll get rid of him and burn [his] goddamned car!" She said, "Don't you dare. Don't you, no, no, no, please!" I'd have the boys do it. You know what? Them boys . . . and all I'd do is buy them a meal once in a while. They never wanted any money. They were good buddies of mine. They were just two rough bastards! I'm telling you, they didn't give a goddamn for nobody. They were really rough guys.[25]

At times it seemed that such tactics were not completely unwarranted. As Minnie Caputo explained, some factory owners were abusive:

> Verbally, they would abuse the women. They were tough, you know, with their mouths and threats. They weren't easy to work for, these guys. They treated us like we were nothing. We couldn't complain. We had to do what they said. They were tough. They were rotten. Couldn't talk to them. They wouldn't listen if you tried to explain anything to them. [Later] when we got the union, we didn't have to talk to them at all. We just talked to our chairlady and the union took care of them, which is the way it should be.[26]

Min's resolve often brought retaliation. Jennie Silverman recalled one incident: "The very worst crisis she had was when they said they were going to come and burn her house down. She sent the kids off. They sent the kids off to relatives and they were up all night, Min and Bill, for a couple of nights to see what might happen. Nothing did."[27]

Betty Greenberg recalled the threats of violence and intimidation:

> One time she was having a terrible time in Pittston. There was a bomb scare. And we went to a friend's house. My mother said, "You've gotta get out of the house, kids, there could be problems."
>
> [In 1958] they threw a red paint bomb at the house. They certainly could have killed her, could have been a real bomb. We were in the house asleep. We didn't even hear it. They threw it outside. The red paint went all over the house, but it could have been a real bomb. We had to get someone to take all this off the house! It went on the neighbor's house. You should have heard them. They took up a petition [that] we should be moved out of the neighborhood! It never went anywhere.[28]

Mounting Successes

By the late 1940s and early 1950s, the ILGWU's campaign was gaining momentum . In 1949 Min rallied Pittston Apparel Company workers, who had not yet been unionized, to strike unless the employer agreed to a contract with the ILGWU. The firm's management and the union negotiated a settlement that included a closed shop and wage increases. With this achievement in hand, the General Executive Board reported that "in the Pittston District, progress has been slow but, recently, union sentiment has been gaining."[29]

In early 1951, Min filed a formal complaint with the Pennsylvania Department of Labor and Industry's Bureau of Mediation against Connie Lee, Inc., a dress and children's clothing manufacturer in Wilkes-Barre. The bureau, established by legislation in 1913, was charged with mediating disputes between employers and labor. As 1950 drew to a close, Connie Lee's management refused to live up to the terms of a contract requiring a 6 percent cost-of-living increase and payment for the Christmas and New Year holidays.

According to the union's complaint, when the shop chairlady questioned management on the matter, she was so severely assaulted that she had to be hospitalized. Thirty ILGWU members immediately walked off the job in protest. Min telegraphed Benjamin Weigand, director of the bureau in Harrisburg, desperately requesting assignment of a mediator. Weigand responded by dispatching P. F. Bolen to Wilkes-Barre. Bolen's inquiry revealed that Connie Lee had indeed violated its earlier agreement. With the threat of legal action looming before the Pennsylvania Labor Relations Board, management engaged in discussions with Bolen and agreed to implement a 6 percent raise immediately and provide holiday compensation. The union celebrated an important victory. Its contract was upheld and members were compensated pursuant to a binding agreement with the employer.[30]

Major organizing drives in the early 1950s yielded further results. In 1951 the union filed a complaint and requested a mediator's involvement in a dispute with Pioneer Manufacturing Company. The Pioneer case and subsequent decisions proved to have an even greater impact than the Con-

A work stoppage in the Wyoming Valley, late 1940s. (Courtesy of Stephen N. Lukasik, Lukasik Studio.)

nie Lee affair. Pioneer employed several hundred unionized workers in its Wilkes-Barre apparel factory. On August 31 the company fired three ILGWU members for insubordination. The workers claimed that they were dismissed for disagreeing with management. Within minutes of the firing, over 200 union members walked off the job and staged an impromptu rally at the front door of the factory. Min requested that Mr. Bolen once again mediate the dispute.

Bolen suggested arbitration, as the facts were much more complex than in the Connie Lee incident. The union alleged that the company had no substantive grounds for firing the three workers and that management had engaged in a lockout and was discriminating against employees who had walked off the job. Pioneer's management argued that it had the right to fire any employees and hire replacements and that the sympathy strikers had, essentially, given up their jobs. The arbitrator's ruling favored the ILGWU's position by calling for the reinstatement of the sympathy strikers as well as of the three workers who were initially fired. The company and union were also required to establish and adhere to a grievance process. The victory further enhanced the union's reputation and viability.[31]

In early 1952 the ILGWU won a substantial wage increase for several hundred members at Leslie Fay, Inc., a company that was fast becoming one of the nation's premier manufacturers of petite women's apparel. Founded in 1946 in Plymouth, Pennsylvania, by the New Yorker Fred Pomerantz, the company expanded its production facilities into several Wyoming Valley communities. The ILGWU organized Leslie Fay workers in the late 1940s, and until the early 1990s the union would enjoy relatively good relations with management. After the Leslie Fay victory, the union added several hundred new members when it successfully organized the Del Mar Dress Company in Pittston, the Steingut Dress Company in Du Pont, the C & S Manufacturing Company in West Pittston, and several small dress and children's wear manufacturers. In total, 1,200 new members were added to the district's roster in 1952, bringing membership at year end to 5,500 in thirty-one Wyoming Valley garment factories.[32]

In 1953 the union's Northeast Department launched a drive to institute a 35-hour workweek, secure wage increases, and maintain an even flow of work from New York jobbers, some of whom had begun to contract with or set up factories in the South. The union's move drew sharp resistance from employers. When its contract with the Wyoming Valley–based Emkay Manufacturing Company expired on July 1, a nine-week strike ensued over wage and hour issues. At the same time workers at two Emkay plants in Kulpmont, Northumberland County (about fifty miles to the south), voted in favor of union recognition and asked the ILGWU to serve as their bargaining agent. Striking Wilkes-Barre workers became involved in the Kulpmont organizing drive to stop Emkay from playing shop against shop in order to drive down wages.

Faced with a long-term disruption of production and a favorable vote for union recognition, Emkay's management agreed to meet with David Dubinsky, David Gingold, and Min Matheson at the ILGWU's Unity House

Wyoming Valley garment workers strike the Emkay Manufacturing Company,
1953. (Courtesy of ILGWU.)

resort in the Pocono Mountains on September 8 to negotiate a settlement.
At the end of the day-long conference an agreement was announced that
reduced the workweek from 40 to 37 1/2 hours. In addition, management
granted a 6 percent wage increase to piece-rate workers, who included all
of the company's sewing machine operators. Wage employees, or those
who earned hourly pay instead of piece rates, were guaranteed the same
rate of pay for 37 1/2 hours as they had received for a 40-hour week.
Finally, the agreement guaranteed a workweek of 35 hours effective January 1, 1956. Louis Rona, manager of the Shamokin District, joined with Min
Matheson to finalize implementation of the contract with Mac Kahn,
owner of Emkay. Within a week over 1,000 Emkay workers were back on
the job.[33]

In 1954 the ILGWU renewed contracts with the Plains Manufacturing
Company, the Glen Lyon Manufacturing Company, and Oestreicher and
Company of Wilkes-Barre. The agreements granted several hundred members pay raises ranging from 15 to 21 percent and reductions in work
hours.[34] In 1955, 250 workers secured ILGWU contracts at the Gluken Bra and

Engle Bra companies in Glen Lyon, the Harsey Blouse Company in Wanamie, and the women's blouse division of Pittston Sportswear. In addition, the Pennsylvania Supreme Court upheld the right of ILGWU members to picket employers as part of organizing campaigns. The decision ended a five-year legal battle that began when a lower court issued an injunction against the union for picketing at Wilkes Sportswear in 1949.[35]

At nearly the same time the union received a favorable decision from the Commonwealth of Pennsylvania's Unemployment Compensation Review Board, which ruled that ILGWU members were eligible for unemployment compensation even though the union paid benefits to members during periods of idleness. The decision settled a dispute that arose in 1953 when several ILGWU members filed for unemployment compensation when they were temporarily laid off. The union's health and welfare fund issued benefit checks to the employees to help sustain their families. In reviewing their claims, the Department of Labor and Industry's Unemployment Compensation Bureau ruled that the workers had received "vacation pay" from the ILGWU and were therefore not entitled to state unemployment compensation. Recognizing that employment in the garment industry is sometimes irregular and sporadic, the Review Board overturned the bureau's ruling and concluded that employees should not be penalized because they receive a union benefit while out of work. The claimants were awarded payment. The ILGWU hailed the decision.[36]

Despite the fact that imported Japanese-made scarves had suddenly appeared in U.S. apparel marketplaces, the General Executive Board reported that the ILGWU's 1953 membership of 445,000 was strong and growing. Impressive gains in Pennsylvania locales—including the anthracite region and its Wyoming Valley—prompted the board to report optimistically that the ILGWU's presence in the Keystone State appeared secure.[37]

Certain events, however, painted a more troubling picture. The May 1949 murder of Min's brother William Lurye in New York made it clear that organizing garment jobbers and contractors remained a difficult and dangerous task.

A Pittston garment worker and her daughter at the Local 295 office, Main Street, Pittston, ca. 1957. (Courtesy of Stephen N. Lukasik, Lukasik Studio.)

The Murder of William Lurye

Like his older sister, William Lurye (commonly referred to as Will or Willie) followed in the footsteps of their father, Max, by becoming deeply involved in the labor movement. Will joined the ILGWU in New York, worked as a dress presser in a Manhattan garment factory, and served as a member of the executive board of Dress Pressers Local 60. He, too, was inspired by the move to bring contractors into the ILGWU fold and volunteered as a temporary organizer when the Dress Joint Board launched a new initiative against nonunion jobbers and contractors in the spring of 1949.

On the sunny spring day of May 9, 1949, the 35-year-old father of four entered the lobby of a building at 224 West 35th Street, near Seventh Avenue in the heart of Manhattan's garment district, to place a telephone call. As he stepped into a booth and fumbled in his pocket for change, three men later described as "tough-looking characters" entered the lobby. Two of the men stood watch while the third forced open the door to the telephone booth and plunged an ice pick deep into Will's chest. As the men fled, Lurye fell to the ground, bleeding profusely. He stumbled out to the street, where two women garment workers came to his aid and summoned assistance. Will Lurye passed away early the next morning at New York's St. Vincent's Hospital.[38]

The murder shocked Manhattan's garment district. Upon receiving word of the tragedy, Dubinsky sent a telegram to Lurye's wife and four young sons:

> With bowed heads and hearts filled with sadness I express profound sorrow in my own behalf and in behalf of our union for the brutal attack, which resulted in the death this morning of your husband and our co-worker.
>
> We realize that very little can be said to comfort you and assuage your grief. But you may rest assured that we will do everything within our power to apprehend the murderer and bring him and his principals to justice. Your husband died in the line of duty as a faithful, loyal and conscientious worker for our union and for better standards and a happier life for all workers. This tragic loss you and

William Lurye shortly before he
was murdered. (Courtesy of
Bernard and Harriet Lurye.)

your family have sustained is irreparable but it may be some con-
solation to know that it is shared by thousands of garment workers
who are filled with great resentment.[39]

Members of the ILGWU in Wilkes-Barre issued a statement expressing
sympathy to Min Matheson and the Lurye family. Noting that "the organizing campaign which William Lurye was engaged in was just another part
of the organization work that we are doing here," local garment workers
assured Min that "neither lies, nor arrests . . . will stop us any more than
hired murderers will stop our New York brothers and sisters in the fight for
decent wages and to banish the chiseling nonunion, low paying shops
from our industry."[40]

At Will's funeral the hearse was led through the heart of Manhattan's
garment district by Dubinsky, representatives of the General Executive
Board, the New York Dress Joint Board, the Lurye family, and tens of thousands of Manhattan's garment workers, who were instructed by the union

to leave their jobs at 10 A.M. to attend the funeral service. New York City police estimated the crowd outside the Manhattan Center, at 34th Street between Eighth and Ninth Avenues—the site of the service—at 100,000 persons. The busiest garment district in the world was brought to a stand-still for several hours as mourners heard numerous tributes to the deceased, including Dubinsky's eulogy, broadcast over a makeshift out-door loudspeaker system. The union president spoke of the ongoing strug-gle for justice in an unjust world. He applauded the commitment to the struggles of working people as exemplified by William Lurye. And, he promised, even murder would not stop the ILGWU's efforts to organize nonunion garment factories.[41]

Deeply distressed, 72-year-old Max Lurye attended the service, as did his estranged wife, Anna, and all of the Lurye siblings and their families. The stress of these events proved too much for Max. A few days after Will was laid to rest, he succumbed. The family claimed his death was due to a "broken heart." Max and Will were buried side by side in Brooklyn's Workmen's Circle cemetery. Their gravestone is inscribed: "With devotion and courage they lived and died for the cause of Labor." The murder of her brother and the death of her father overwhelmed Min Matheson. In such despair that a simple request to pass a butter knife at the dinner table plunged her into deep melancholy, Min withdrew from union affairs for a short time.[42]

Attention immediately focused on apprehending the murderers. The union posted a $25,000 reward for information leading to their arrest. Police investigations and eyewitness accounts identified two of the mur-derers as Benedicto Macri and John Guisto. Macri's sordid career included stints as an officer of a New York ship repair company, part owner of a gar-ment factory, and partner in a nonunion trucking company. His business associations linked him to Albert Anastasia, New York's crime boss and head of the so-called Murder, Inc. Anastasia was part owner of the ship repair company, provided capital for the venture, and personally hired Macri. Anastasia also owned a nonunion garment factory in Hazleton, Pennsylvania, about twenty-five miles south of the Wyoming Valley, and was godfather to Macri's child.

Guisto, an ex-convict employed by Macri in the garment business as

The Lurye family at the funeral of William Lurye, Manhattan Center, New York, May 1949. (Courtesy of Kheel Center Archives, Cornell University.)

well as in illegitimate activities, boasted a long list of criminal activities, including armed robbery and larceny. The third accomplice remained unidentified. By June 1949, both men were indicted for the Lurye murder. Despite their positive identification, they remained at large for a year. The *New York Post* reported that "New York has not forgotten. The case will not be closed until the murderers are caught. There can be no real rest or refuge for the killers because New York will remember Will Lurye."[43]

In June 1950, Macri surrendered to police; Guisto remained at large and was never apprehended. Macri's trial began in October 1951. The prosecution argued that Lurye was murdered by Macri and that Guisto was an accomplice. Though indictments were never issued against any specific employer, the prosecution argued that Lurye's murder was most likely motivated by Lurye's efforts to organize at least one contractor in New York. It was also suspected that certain nonunion jobbers had hired the killers to send a warning message to the union and its organizers. The prosecution's case collapsed when two eyewitnesses suddenly changed their testimony: they could no longer positively identify Macri. The trial ended in acquittal.[44]

Maxine Lurye, Min's sister, accused Dubinsky of failing to insist that all avenues be vigorously pursued to secure Macri's conviction. Dubinsky should have used his power to encourage the authorities to prosecute Macri on charges of perjury, she argued. She also recommended that the ILGWU pursue the employers who had hired the killers of her brother:

> It has been some months since the freedom of Macri. The Lurye Family has been dealt one blow after another. My poor Dad. No wonder he died of a heart attack. He knew how deep this betrayal was. For him, Mr. Dubinsky, I personally will not remain quiet and keep holding the fort that maybe something will happen to bring justice to us all. You, Mr. President, can do something. Do it! Keep your promise.
>
> All my mother's life was spent in hunger, poverty and misery to make it possible for her husband and her children to contribute towards the growth and power of the labor movement. Is not my mother entitled to feel that her suffering was not in vain? Will this go unanswered too?[45]

Embittered by the whole affair, Will Lurye's mother appealed to Governor Thomas E. Dewey of New York to assist in the quest for justice. Min likewise pleaded with Manhattan's district attorney, making it clear that in the family's opinion, both Lurye's murder and the sudden change of eye-witnesses' accounts were "sponsored and nurtured by [Albert] Anastasia" and that "if ever there was a flagrant miscarriage of justice it was that pretense of a trial accorded Benedicto Macri for the murder of William Lurye."[46] Despite such appeals, the case was not resurrected. No one was ever convicted of the murder.

A free man, Macri returned to the streets of Manhattan. In the spring of 1954, he was knifed to death on a Lower East Side street. Speculation immediately attributed his death to his involvement in the Lurye affair—vengeance, perhaps. While the underlying reason for his murder and the actual culprits would never be known, one assumption was that the ILGWU and perhaps the Lurye family were somehow involved. Such speculation was so widespread that the American Broadcasting Company's Walter Winchell hypothesized during his live broadcast on Sunday, May 2, that "the police have a new tip on the murder of Macri—garment unions in and near Wilkes-Barre, Pennsylvania. It may be vengeance for the murder of Lurie [*sic*] a few years ago, the New York union organizer."[47]

Though Min denied any involvement in Macri's murder, the police focused attention on her and her siblings. "Last night I was visited by two representatives of the New York Police Department," Min wrote to Dubinsky. "They came with a Sergeant Peck of the Penna. State Police. They were investigating the possibility that I or some members of my family had something to do with the murder of Benedicto Macri. They spent over three hours in my home and then asked for addresses of the rest of my family."[48] No evidence linked the Luryes or the ILGWU with the murder. Macri's death remained unsolved.

The fact that she had lost her brother and father—people who shared deeply in her union philosophy—reopened deep psychological wounds for Min Matheson. The association of Macri with underworld figures such as Albert Anastasia strangely harked back to a time when Max Lurye nearly lost his life at the hands of Capone's assassins in Chicago.

For her, once again, individuals of questionable character and with dubious backgrounds were standing in the way of progress for working people. These individuals and the criminal elements they represented epitomized evil and injustice. And they would go so far as to murder and sway witnesses to protect their rackets. Will's murder was particularly complex for Min because he gave up his life protesting the circumstances for which she, a hundred-odd miles away, was seeking redress. She was organizing nonunion contractors who received the bulk of their work from New York jobbers, some of whom her brother had struck and who perhaps—though it was impossible to determine—had hired his killers.

Yet, rather than causing Min to recoil in fear of the same fate, Will's murder fueled her over the long haul to fight for the ILGWU's cause. Sol Hoffman, an organizer who befriended her in the 1950s, recalled:

Because her brother was killed, Min had a personal stake in doing what she did. She took them [organized criminals] on. She had a personal stake to really go after them. She had a determination and was successful. In this union Min was recognized as a leader in going after them, especially in the '58 strike.

I remember hearing her in Pittston, "Bufalino this" and "Bufalino that." Publicly she would say such things. I was new to all of this at the time. But I remember thinking, "My God, this woman has a lot of guts!"[49]

With the assistance of her husband and a core of organizers, Min continued to organize and worked to establish an infrastructure that would serve the needs of the ILGWU's growing membership.

4

Building a Union Infrastructure

To infuse into masses of people, still new to this organization, a love and
devotion to their union [has] required a genuine mass education effort in addition
to the very essential gains of improved working and living conditions.
—*Report of the General Executive Board of the ILGWU to the Twenty-fourth Convention*, 1940

We cannot forget the poverty, the sickness, the homework shops, the child laborers.
Life in this nation can be something better than a slum.
—David Dubinsky, 1955

In 1946 the ILGWU's General Executive Board appointed Min Matheson as
manager of the newly created Wyoming Valley District. Bill Matheson
assumed the position of education director. The district was divided into
three locals: Wilkes-Barre (No. 249), Pittston (No. 295), and Nanticoke (No.
327). By 1950 membership topped 3,000 and its continued growth provided
Min and Bill with the opportunity to implement the ILGWU's commitment
to "social unionism," the idea that a labor union ought to have an agenda
beyond organizing workers, winning higher wages, and securing
employer contracts. At least part of the ILGWU's mission included building

an infrastructure that engaged worker-members in social and community activities, provided for their needs, assisted them in times of personal difficulty, exposed them to education and culture, and involved them as activists in social and political causes. The idea was to expose garment workers to a union "way of life" and to a culture that valued mutual association and camaraderie. While she continued to organize, Min increasingly relied on her husband to assist in building an infrastructure that would enable the union to consolidate its position in the community and become something beyond a pay-and-benefits organization.

Initially the infrastructure consisted of three components. First, Bill created a district newsletter, *Needlepoint*, which served as the main source of union news, information, and advocacy. Second, the district operated a health care center in Wilkes-Barre, the only one of its kind in the anthracite region. And third, the union created a highly successful and popular chorus.

Needlepoint

Not long after Min and Bill arrived in the Wyoming Valley it became apparent that the ILGWU needed a way to communicate reliably with union members. Consequently, Bill spearheaded the production and distribution of a periodical newsletter. He gave it a title to which all garment workers could relate: *Needlepoint.*

Very much the district's intellectual leader, Bill possessed exceptional research, writing, and literary skills, and, according to Betty Greenberg, channeled his diverse interests in a variety of directions:

> My father knew reams and reams of poetry by heart. Shakespeare, Longfellow, Whitman, Bobby Burns. He loved Lincoln's writings. He knew history. He used to always say to my sister and I, "Don't you want to know this to know it? Don't you just love learning, because learning is so wonderful?"
>
> My father believed [you] don't say anything is wrong until you know the facts. When you know the facts, you have a right to say it is wrong. Don't prejudge anything until you know the facts. Most

people don't know anything about the *Communist Manifesto*. They just know that communism is no good. Well, communism is no good because of what people did with it, not because of what it stood for. My father had a way of saying that too. He would say, "Against communism, huh? Do you think that every man should have a right to make a decent living? Be able to feed his family? Not have to worry about where his next meal is coming from? Or about having a roof over his head? Do you feel that every man should have a right . . . " and he would go through different things. Then he would say, "Well, by God, I guess you believe in communism."

My father was an atheist. He brought us up atheist. But my father could quote the Bible very, very well. We would have to know the Bible. You have to understand anything that you were going to say is not true or right. You would have to know about it first.

Along with his intellectual abilities, Bill displayed wide practical and literary talents:

My father was the cook in the house. He did all the cooking. Oh, my God, he was a good cook! He could do everything. Fix plumbing, the lights, the vacuum cleaner. And write wonderful literature! Speeches. He always helped us with our work in school. We never got anything but an A in anything that he helped us with, even in college.

He had such ability to write. My mother used to get angry with him. She used to tell him that he should have written the story of their lives. But finally, when he was humble enough that he was thinking about it, he became senile. He should have been a professor. But he viewed himself as just another worker. "We are all equal. There should be no separation," was what he would always say. It was just in my father. All his life he was like that.[1]

Jennie Silverman remembered that folklore stood out among Bill's many intellectual interests: "He was a sweet, gentle person. He wrote poetry and was a storyteller. He'd get lost in it and stay up past midnight. He was such a fantastic storyteller. They were sort of homey stories. Probably things he himself had lived through. It wasn't stories he had read in a book. How often does a guy recite Keats by the yard? That was Bill."[2]

When ILGWU headquarters appointed Bill education director for the Wyoming Valley District, he dedicated his efforts to aiding Min and other organizers in several ways. He wrote speeches for Min and helped to strategize for pickets, protests, and public events. He also took responsibility for longer-term initiatives such as creating a workers' education program.

The first issue of *Needlepoint* appeared as a mimeographed information bulletin in December 1945. It consisted of a few pages that explained the ILGWU and ways in which it sought to improve workers' lives. It espoused the union's view that it was the right of every worker to expect decent wages and working conditions, overtime pay, vacation benefits, and assistance in time of sickness. As district membership grew through the late 1940s and 1950s, so did *Needlepoint.* By 1951 Bill produced the newsletter quarterly and by the mid-1950s monthly. *Needlepoint* included several special features. Min wrote "Our Manager's Column" to discuss issues ranging from local strikes and work stoppages to organizing efforts to the union's views on politics and public policies. Various articles discussed political issues, candidates, elections, voter registration, political endorsements, strikes, organizing drives, and shop floor matters. A section headed "Proverbs" contained poetry and short stories about life, work, family, and community. Discussion of local events, as well as the ILGWU's involvement in them, also became a regular feature.[3]

To provide workers with a voice, Bill started a letters-to-the-editor section. He also included discussion of union-sponsored education programs as well as an index of books and other educational materials in the district's library, located at its headquarters in Wilkes-Barre. The newsletter regularly reviewed the status of state and federal legislation that affected working people. Members, union leaders, candidates, and political officials wrote guest columns. *Needlepoint* also included notices on union benefits and announcements of births and deaths.[4]

By 1963 *Needlepoint* was distributed to all of the nearly 10,000 district members. It furnished a direct link between union leaders and rank-and-file members while providing both with a means to explore their mutual interests and to connect with the larger organization. *Needlepoint's* value was recognized by the ILGWU's Wyoming Valley District Council:

> In 1945, Wilfred Matheson, working as an organizer and Educational Director, issued a mimeographed paper called *Needlepoint*. This little paper, now printed, has given our members the courage to organize and join the union and has educated the membership. Today, this paper is the most powerful and popular method we have to reach the members of the union. We desire it placed in the records that we request that Bill Matheson issue *Needlepoint* regularly for the Wyoming Valley District no matter where he be located. This has been resolved and so be it recorded, this day, Monday, January 28, 1963.[5]

A Garment Workers' Health Care Center

Long before Min and Bill Matheson had joined the ILGWU the union worked to address the poor health conditions prevalent among New York's garment workers. Inadequate diet, tuberculosis, anemia, and other afflictions were common among those who labored in the sweatshops of the Lower East Side:

> There are no vital statistics to show the rate of mortality and morbidity among cloak workers. The general impression, however . . . is that they are not of a robust type, that they largely suffer from anemia, possess a more or less stooping gait, and are not, as a rule, a very healthy lot of men and women. There are no definite figures as to the extent of tuberculosis among cloak workers, but it is well known among the physicians of the East Side that tuberculosis is a very frequent disease among this class of workers.[6]

In 1910 ILGWU convention delegates and the General Executive Board agreed that physical afflictions among garment workers were as substantive a concern as consumers' purchase of apparel manufactured in unsanitary environments. Public health, the union argued, was at risk, since the epidemiology of diseases such as tuberculosis was not yet fully understood. The union addressed this issue in 1925 when it introduced the "Prosanis label"—or a union label—attached to garments produced in factories that met certain health and environmental standards.[7]

A 1915 ILGWU-supported study of New York garment workers documented widespread maladies:

> Tuberculosis was, undoubtedly, the most important disease among garment workers. Three and eleven hundredths percent of the males examined and nine-tenths of one percent of the females were found to be tuberculous. This is a rate of prevalence for females of nearly three times and among males nearly ten times that of this disease among soldiers in the United States Army, for instance. In many instances the subject was unaware of his or her condition, having been conscious only of general impairment of health. Tuberculosis is unduly prevalent among garment workers, especially among males. A faulty posture was extremely common among garment workers.[8]

One outcome of the Protocol of Peace was the establishment of the Joint Board of Sanitary Control, consisting of union, employer, and public representatives. The board established, implemented, and monitored sanitary conditions in Manhattan garment factories. The board also created and sponsored the Union Health Center, established in 1914. The center made available examinations, screening and testing, treatment, and health education classes to garment workers and their families at little or no charge. Within a few months of its opening, the center provided services to some 3,000 workers. In 1919 it relocated to larger quarters in New York's Union Square and was soon recognized as a model for provision of health care to working people. On a visit to commemorate the center's golden jubilee in 1969, President Lyndon Johnson commended the ILGWU for its progressive leadership in health care:

Against the bitter obstinacy of entrenched interests they [ILGWU] battled, first to free workers from the slavery of the sweatshops, then to free them from sickness and disease. This health center is a testimony of their success and a memorial to their spirit. Fifty years ago, this health center stood alone, the first of its kind to be established in our country by a trade union for working men and working women. Your union stood resolute in the thin ranks of those who carried on the struggle for security for the helpless, who fought the battle for a better life for every citizen.[9]

As the apparel industry and the ILGWU moved beyond the geographic horizons of New York, so did the union's concern for the health and welfare of its members. By the time of President Johnson's commemoration, the ILGWU operated health care centers in over a dozen districts throughout the United States. Pennsylvania's centers were established in Philadelphia, Allentown, and Wilkes-Barre in addition to a mobile unit that traveled the state.

The Wyoming Valley District's health care center opened in June 1948. It served garment workers and their families throughout the anthracite region. The center was funded by employers' contributions to a health and welfare fund administered by the ILGWU's Northeast Department. Such contributions had become the norm in contracts negotiated with employers in the post–World War II period. David Dubinsky chaired the center's inauguration ceremony, which was attended by union officials, community members, local politicians, rank-and-file members, some garment factory owners, and the press. The *Scranton Times* reported:

> A luncheon attended by important union officers, employer representatives and community leaders, and a "Home Town" fashion show will mark the formal opening Friday and Saturday of a $150,000 Union Health Center in Wilkes-Barre to serve 7,000 members of the International Ladies' Garment Workers' Union in the Wilkes-Barre, Hazleton and Scranton areas.
>
> Representatives of the employers and the union will meet at lunch with community leaders and health officers at the Hotel Red-

ington, Wilkes-Barre. Members of the union's general executive board, gathering from coast to coast for their quarterly meeting, will attend. Following the luncheon, guests will make a tour of inspection of the health facility. Services will be exclusively diagnostic with no charges to ILGWU members. The center is financed by employer payments into the union's health and welfare funds, as negotiated in union contracts.[10]

Min championed the center from its inception:

> Almost immediately we opened a health center for our people. I started it here in Wilkes-Barre in 1948. And we still have it on Washington Street. It's still there. It was an old furniture store, and of course we [re]built it. Some local doctors were against it because they felt it was socialized medicine.
>
> We had the finest set of doctors and we tried in every way, you know, to help them maintain their health. You could get a cardiogram and you could get blood work done. The proctologist is available because cancer is so prevalent. If you needed medication, we had a prescription center. All the important tests were done in the health center free to the members.[11]

Health services included routine physical examinations, X rays, cancer screening, electrocardiograms, allergy tests, psychiatric consultations, proctology examinations, ear, nose, and throat diagnostic services, immunizations, and gynecological services. Health education classes and outreach programs were important parts of the center's activities. Workers received instruction on the importance of twice-yearly examinations for detection and prevention of cervical and breast cancer. Literature warned against fad diets and provided information on nutrition, fitness, vitamins, and dietary supplements. The center also educated members on the importance of preventive health care: "The earlier a disease is discovered, the easier and quicker the cure. Have you had a medical examination? It's free for the asking at your UNION HEALTH CENTER. Phone VAlley 3-6127 in Wilkes-Barre."[12]

Immunizations at the ILGWU Health Care Center, 1950s. (Courtesy of ILGWU.)

Bill Matheson's literary skills contributed a personal touch to encourage members' use of the facility, particularly when it came to women's health problems such as breast cancer: "Don't wait until it's too late! Your doctor will tell you it's best not to pass up a lump in your breast. Though it may be benign, it's a clear warning sign. Call your health center to give it a test!"[13]

The union's goal was to have each member receive at least one complete physical examination annually. In its first five years the center provided over 123,000 tests and services. By 1958 over 20,000 union members and their families from Scranton, Wilkes-Barre, Hazleton, Pottsville, and other anthracite communities had received over 220,000 tests and services.[14] In commemorating the eleventh anniversary of the Center, Dr. Albert Feinberg, its director, remarked on the significance of its work:

Let us reflect, for a moment, on the events that have transpired in the past 11 years. It took vision on the part of union officials to see the possibilities of such a revolutionary undertaking in northeastern Pennsylvania. A revolutionary venture. I say revolutionary, because this type of medical care was something entirely foreign to what we of the medical profession thought was the best and only way to practice medicine.

The physical plant of the Health Center has grown three-fold from its small beginning. The staff, both medical and technical, have been increased many times. The laboratory and clinical services made available to patients attending the center is second to no other out-patient clinic anywhere. Our Health Center was the first in northeastern Pennsylvania to do mass cervical biopsies and Papanicolaou stains in the Union's cancer detection program.

Men come and men go. But the good services of the Union Health Center of the ILGWU will go on and on. It will remain an ornament to the cooperation of organized labor and industry, a glory to the community and a benefactor to the human race.[15]

With *Needlepoint* as a vehicle to reach garment workers and with the Health Care Center to address their physical needs, the district had put into place two components essential to establishing the ILGWU's presence in the community. Min and Bill also worked to build the self-esteem, morale, and community image of workers by creating a popular and widely recognized chorus.

The Wyoming Valley District Chorus

Needlepoint and the health care center had successfully demonstrated that the ILGWU did more than arouse workers regarding conditions in the workplace. Min and Bill's decision to create a chorus—modeled on a New York-based ILGWU theater performance group—was another measure of their commitment to shaping the district's infrastructure.

Tapping into the artistic creativity of its members was nothing new to the ILGWU. In the mid-1930s the union purchased and operated the Labor Stage Theatre in midtown Manhattan and assembled a troupe of performers drawn from the rank and file. The group debuted with a play titled *Steel*, which dramatized life in an American steel town. The play coincided with the efforts of the CIO-backed Steelworkers' Organizing Committee (which would later evolve into the United Steelworkers of America) to unionize the steel industry. In 1938 the group produced and performed a musical known as *Pins and Needles*, depicting historical and contemporary aspects of the American labor movement and working-class life.

The popularity of *Pins and Needles* grew quickly as it received accolades from the labor movement and the New York theater community. Two traveling companies were formed and Hollywood performances were booked. One company was invited to perform in the East Room of the White House for President Roosevelt and other Washington officials. Bolstered by enthusiastic reviews and a high-profile performance in the nation's capital, by the early 1940s *Pins and Needles* had become the longest-running Broadway musical up to that time. It launched the ILGWU's long-term commitment to entertainment as a vehicle to espouse the causes of working people and the labor movement.[16]

The Mathesons believed it was important to engage Wyoming Valley workers in the type of cultural and artistic projects that had become an ILGWU hallmark. They also recognized the value of providing social outlets to boost esprit de corps and promote a positive union image. Min and Bill recruited the local members Jim Corbett, Bill Gable, and Clementine Lyons to organize a Wyoming Valley version of the New York troupe. According to Clem Lyons,

> In 1947 we got the shows going. We were, in fact, the first district in the Northeast Department to put together a chorus and put on a show. New York, of course, had the Labor Stage. We soon followed.
>
> Well, in 1947 Min Matheson came to me to try and find people to go to the first meeting for the chorus. And you'd go into the shops and sit beside them on the lunch hour and coax them out. But it was very difficult to get them interested because, in some cases, they were afraid. And their husbands wouldn't let them participate.

There were some that, once they came out, they were just happy to come out. So I got about eight or ten to go to the meeting, and sure enough, they put on the [first] show down at the Kingston Armory. And Jim Corbett had me do a number.

In 1953 Mrs. Matheson wanted to put on a kiddie show. I took on the first kiddie show. We called it the "Lollipop Revue." So in 1954, when it came time to go into rehearsals for the regular annual musical, Mrs. Matheson let me and Billy Gable do the shows for her. Every year from 1947 to 1976. I had been in them from 1947.

Billy Gable was the musical director. He worked in the office in Wilkes-Barre. We'd hire a dance instructor and we could only afford a dance instructor once a month, so I'd teach them dancing too. Well, first of all, they thought that it might be a waste of time and we'd really have to take them by the hand and say, "Look we'll meet you in the union office," and provide rides for them or go and pick them up at their houses. Billy Gable and I would be working, dogging people to come out and take part in the activities.

According to Clem, before long the popularity of the chorus became apparent:

In fact, I'll tell you, not that I want to brag because I was part of the show, but the shows began to be a part of the social calendar around the area. People said, "When are they going to have another show"? Or "When are you going to have tickets for me?" Or "Is the gang from the ILG going to be there? We'll buy tickets." You know, that kind of thing. In fact one year we sent $450—at that time $450 was a lot of money, I think they charged 50 cents to get in to see the show—and then we sent the money up to St. Michael's [a local school for orphans] to buy athletic equipment for the kids up there.

Some [performances] would be, maybe, popular shows that were on Broadway at the time, but most of them would be union flavor. We used to go to the Veterans' Hospital [in Wilkes-Barre] every Thursday night for years. Around '52, I think, we helped the Ameri-

Chorus practice, 1954. Standing on the right is Clementine Lyons. (Courtesy of Stephen N. Lukasik, Lukasik Studio.)

can Italian Association with their show. [We performed] at the American Legion in Exeter [near Pittston]. [We had] a medley, "I Could Have Danced All Night," "Darkness on the Delta," "Buttons and Bows." We had maybe forty-eight people in that group.[17]

The first large-scale musical narrative performed by the chorus was *My Name Is Mary Brown*. It debuted in 1950 and featured original music and lyrics written by Michael Johnson, an ILGWU organizer, and Jim Corbett, co-director of the chorus, with the assistance of Bill Matheson and Leon Stein from New York. An opening solo introduced Mary Brown, a garment worker employed in a runaway factory:

My Name Is Mary Brown

I am Mary Brown.
I come from Pennsylvania,
Vermont and Massachusetts.
New Hampshire and Rhode Island.

From Delaware and New York State,
From Maine and from New Jersey.
My Name is Mary Brown.[18]

The musical married history with the contemporary state of affairs in the U.S. garment industry to tell the story of how Mary Brown, like New York garment workers of an earlier era, worked long hours in horrid conditions in a nonunion factory. Songs such as "Won't You Come Along" and "This Is a Strike!" told how Mary Brown and her co-workers, disillusioned with working conditions, banded together with ILGWU organizers to shut down the runaway factory and demand higher wages and shorter hours. Several songs explained that conditions improved after the successful strike and that membership in the union yielded benefits well beyond the shop floor. "Here's to Your Health" explained how ILGWU members benefited from the services of the union-sponsored health care center. "Up at

Unity House" highlighted the ILGWU's vacation and workers' education center in the Poconos, the only union-owned facility of its size—and one of the few of its type—in the United States:[19]

Up at Unity House

Gee, but it's fun when our work is done,
To be up at Unity House.
Free as the breeze 'mid the flowers and trees
When we're up at Unity House.
We have such fun in the summer
Up in the Pocono Hills.

Gee, but it's great. Oh! You must make a date,
And we'll find such romance and thrills
Up at Unity House.

Some like to go canoeing
While others climb on the rocks.
Some think of nothing but wooing
Others eat bagels and lox.

Some get exercise daily
Dealing a pinochle deck.
And it all comes true when the Union gives you
That annual vacation check!

We have such fun in the summer
Up in the Pocono Hills.
There's so much sport and the week is too short
When you find such romance and thrills
Up at UNITY HOUSE!
Up at UNITY HOUSE!

The musical concluded with "Help Us to Organize!" which called on garment workers to unionize runaway factories and build the ILGWU to ensure better pay, benefits, and working conditions.

My Name Is Mary Brown was performed in numerous communities in Pennsylvania, including Allentown, Scranton, and Philadelphia. Buttressed by performers from other ILGWU districts in the Northeast Department, the chorus traveled to several locales in the northeastern United States where runaway factories had become a problem and where the department had set out to organize, such as Binghamton, New York, and western Massachusetts. *Mary Brown* was also performed at Unity House.

The play provided the chorus with a significant entrée to the larger anthracite region community. The group performed for ethnic and civic clubs, churches, political events (usually for the Democratic Party and Democratic candidates), garment factory parties, hospitals, and community agencies. Holidays were always busy. Memorial Day, Labor Day, Fourth of July, and December–January holiday events brought numerous invitations. By the mid-1950s it was common for the chorus to perform at least once monthly. By the end of the decade performances grew more frequent. In December 1959, for example, the group entertained at sixteen community events in addition to staging performances exclusively for district members and participating in the ILGWU's annual holiday program in New York City.[20]

Initially its productions consisted of melodies from *My Name Is Mary Brown*. By 1952 annual musical revues varied. The programs drew on the history of the American labor movement, the garment industry, and the ILGWU and dealt with themes of social justice, workers' rights, and political activism.

In early 1952 the chorus staged a performance titled *Meet the Girls* in Wilkes-Barre. The revue, with a cast of sixty, featured songs and dances that depicted garment factory work, described the mission of the ILGWU, and linked the union's agenda with the larger goals of the labor movement and progressive politics. The troupe offered a special performance for delegates to the Pennsylvania Federation of Labor's annual convention.[21]

A chorus performance, 1965. (Courtesy of Stephen N. Lukasik, Lukasik Studio.)

The chorus staged a benefit revue titled *Everything Goes* before sell-out crowds at Pittston High School in 1959. The program mixed popular Broadway melodies with union-flavored songs in both solo and group renditions. All of the proceeds were donated to the families of twelve coal miners who lost their lives in the January 1959 Knox Mine Disaster.[22]

In recognition of the ILGWU's twenty-five-year presence in northeastern Pennsylvania (1937–62), Jim Corbett, Bill Gable, Mike Johnson, and the Mathesons created a new program titled *Make Way for Tomorrow*. This revue drew on familiar themes of ILGWU history to demonstrate the union's progress in organizing 450,000 workers nationwide and securing benefits—vacation pay, sick benefits, and a 35-hour workweek. Original songs such as "It's So Different with a Union" championed the ILGWU as working people's best assurance for a future of security and prosperity:

Wyoming Valley District Chorus members in a Labor Day parade, 1960s.
(Courtesy of Stephen N. Lukasik, Lukasik Studio.)

It's So Different with a Union

It's so different since we got the Union,
It's so different than it used to be.
We're half-a-million strong
As we proudly march along.
What a feeling to be free!

It's so different since we got the Union,
Now the days don't seem to be so long.
Everybody's treated fair,
Everybody gets a share—
What a feeling to be strong.

Remembering - ILGWU Christmas Party, 1961

The ILGWU Christmas party for children, Pittston, 1959. (Courtesy of Stephen N. Lukasik, Lukasik Studio.)

No more, no more exploitation,
Hail the freedom we have won—
What a wonderful sensation,
One for all and all for one!

It's so different since we do the Union,
And the things it's done for you and me.
Once we never could be sure,
Now our jobs are all secure.
It's different than it used to be.

[Women]
Before we ever had the Union,
Remember how they used to pay.

ILGWU members join the Salvation Army in organizing assistance for victims of the Knox Mine Disaster, Port Griffith, January 1959. (Courtesy of Stephen N. Lukasik, Lukasik Studio.)

If you dared ask for more,
They would show you to the door.
They can't do that to us today.

[Men]
It's so different now in Pennsylvania,
In New England towns were on the map.
Now, we blast the run-away,
And we really make him pay—
Fighting hard we bridged the gap!

Thirty years has taught the Northeast
That enforcement must be tough.
We have seen our standards increased,
For we always treat 'em rough!

It's so different since we got the Union,
As we march along in unity.
And we want you all to know
That we're gonna keep it so
With the good old ILG![23]

The chorus also became a medium through which the district advocated a larger political agenda. Among its first politically oriented bookings was an invitation from the local Democratic Party to perform on Public Square in Wilkes-Barre during a 1948 welcoming ceremony for President Harry Truman. By the mid-1950s the chorus gained the recognition of the Pennsylvania Democratic Party when it was invited to the state capital to open a rally for Adlai Stevenson, making his second bid for the presidency in 1956: "Thirty members of the Wyoming Valley District of the ILGWU traveled by bus to Harrisburg last night. [The] ILGWU chorus sang at the Democratic rally for Stevenson and [Estes] Kefauver. The chorus sang special political campaign songs: 'We're for Adlai,' 'You Gotta Know the Score,' 'On for Stevenson,' and 'You Must Be There.'"[24]

Bill Matheson worked with chorus members to choreograph the performances and wrote songs for political events. Several of the songs he authored supported local, state, and national politicians friendly to the union's cause, such as Congressman Daniel J. Flood, who represented the Eleventh Congressional District of Pennsylvania from the mid-1950s to 1980:[25]

The Congressman for Luzerne
(Tune: "The Yellow Rose of Texas")

There's a congressman for Luzerne County
Who serves both you and me,

No other one can match him
On that we all agree.
We won with him many times before
And now we're here to say
That Dan goes back to Congress
On next Election Day.

He's the servant of the people,
The best we ever knew,
He works for us from early morn
Until the day is through.
They may talk of other congressmen
In states from sea to sea,
But we have a friend in Dan Flood,
He's the only one for me.

We're going to reelect him
Each time that he will run.
We know we can depend on him
When duty's to be done.
We'll join together gaily
And sing the songs of yore,
The congressman for Luzerne County
Is ours forevermore!

With the assistance of Bill Gable and Jim Corbett, co-directors of the chorus, Bill Matheson wrote original music and lyrics to encourage workers' participation in the electoral process:

Politics

You may read it in the papers,
You may see it on TV,
It may happen down in Nanticoke
Or here in Wilkes-Barre,

Building a Union Infrastructure

It may be up in Du Pont
Or some town along the way.
Your vote is needed next election day.

Politics is everybody's business
Politics is everybody's job,
We've got a job to do
And it's up to me and you.
Politics is everybody's job.

They must register and vote
'Cause one vote here
And one vote there
Can make us miss the boat,
And many of our friends
Have lost by margins very thin.
Then we wound up with nothing
When the other guys got in.

Another important politically oriented song encouraged members to remain informed about public policy and political issues and support selected candidates for public office:

You Gotta Know the Score

Working gals and working guys,
Everyone got to know the score.

Ev'ryone come lend an ear, 'cause
This is an election year.
It's time we all knew where we stood,
Time to pick the bad from the good.
They've had us fooled for quite a while,
But now we're wise to their style.
They're pretty sharp with double talk

But it's time we made a squawk.
We know the bills they voted for—
They're not for us, that's for sure.

Chorus:
You gotta know the score.
You gotta know the score.
So don't give in 'cause you're bound to win
When you know the score.

When you take yourself to a baseball game
You like to know every player's name.
You want to know how he hits the ball,
If he's giving it his all.

You cast your vote on election day,
You want to know how the players play.
You look up his record and there you'll see
If he's on the ball for you and me.

Chorus
So tell your friends and your neighbors that
Every worker must go to bat.
Tell them all to get out and vote.
If you don't. then you're the goat.

If your senator or congressman
Doesn't vote for the working man,
If he doesn't prove that he's labor's friend
Your vote will get him in the end.

We need more supporters for JFK and LBJ,[26]
Men who think of a better day.
Men who'll battle with you and me
To build this great society.

We need better housing, a better wage,
Better schools in this day and age.
Rights for ev'ryone, rights for all.
Join with us and heed the call.

Chorus

Complementing its political overtones, the chorus performed songs to foster solidarity and sustain workers during organizing campaigns, strikes, pickets, and related actions. One was to the familiar tune of "We Shall Not Be Moved":

We Shall Not Be Moved

The union is behind us,
We shall not be moved.
The union is behind us,
We shall not be moved.
Just like a tree that's planted by the water,
We shall not be moved.

We're fighting for our freedom,
We shall not be moved.
We're fighting for our freedom,
We shall not be moved.
Just like a tree that's planted by the water,
We shall not be moved.
Chorus

We're fighting for our children . . .
We'll fight for compensation . . .
We'll build a mighty union . . .

New York ILGWU officials credited the chorus with bolstering the union's community image and promoting a worthwhile and visible agenda in

Pennsylvania. According to David Gingold, vice president of the North-east Department,

> Changes in community attitudes particularly in Pennsylvania were heavily influenced by live revues featuring lyrics, music, sets and acting by members as part of the Northeast Department's educational activities.
>
> Proceeds from numerous performances are contributed to worthy local agencies and causes thus strengthening the union's ties with communities and expanding local respect for the union. Some of those aided at that time were . . . the Hospital for Crippled Children in Wilkes-Barre.[27]

In 1957 over 1,000 area garment workers assembled in the gymnasium of Wilkes-Barre's King's College to celebrate the ILGWU's success in improving their lives and livelihoods. Introduced by David Dubinsky, Governor George M. Leader gave the keynote address, followed by a chorus performance that included a number of political songs that highlighted the ILGWU's commitment to the Democratic Party and its leaders, including the young and popular governor.[28]

For the remainder of the 1950s and the 1960s, the chorus continued to stage annual musical revues. Among the community organizations and institutions it benefited were the Knights of Columbus, Salvation Army, Boy Scouts, March of Dimes, Little League baseball, and local hospitals. The ILGWU's leadership showcased the chorus at the union's holiday festivities and at its triennial conventions, which drew several thousand members to places such as Madison Square Garden and Miami Beach. Chorus members also participated in annual spring fashion parades through New York's garment district. By the mid-1950s the troupe's numerous performances at Unity House had entertained thousands of members of the ILGWU and their families during summer operating seasons. Political bookings continued as well. A few weeks before the 1960 election, the chorus opened a rally on Public Square in Wilkes-Barre when John F. Kennedy came to campaign for the presidency.[29]

The chorus became a key promoter of the ILGWU's widely known "Look

Members of Local 295, Pittston, prepare Easter candy baskets for children, 1959.
(Courtesy of Stephen N. Lukasik, Lukasik Studio.)

for the Union Label." This anthem was adopted as the finale of many performances, and still is:

Look for the Union Label

> Look for the union label, sister,
> Look for it when you buy.
> Tell your brother and your mister—
> Keep a watchful eye.
> It's good for you and it's good for me,
> It's good for all our family.
> So look for the union label when you buy.

Make sure the union label's in
Ev'rything you wear.
If it doesn't have the label, then you
Know the work's unfair.
Let's show the chiseling bosses
The power it can wield.
We'll make that union label
The garment worker's shield.[30]

In 1965 members of the Wyoming Valley District chorus combined their talents with others from the Northeast Department to produce a $33\frac{1}{3}$ RPM record album for the thirty-second convention of the ILGWU, held in Miami Beach. Music and lyrics provided a comprehensive history of the ILGWU from the sweatshops of New York's Lower East Side to its 1965 membership roster of 460,000. The ethnic mix of the membership was reflected in the surnames of the artists who performed live before 1,000 convention delegates: Suriano, Wasko, Weiss, Castiglia, De Annuntis, Trigiana, Pickett, Philips, Pingitore. . . . The chorus closed the show with a finale that boasted of the progress of the ILGWU and the Northeast Department, incorporating a familiar union song that would become another of its trademarks, "Solidarity Forever":[31]

Finale/Solidarity Forever

We're from the Northeast Department—and today
We're proud we're thirty years of age.
Here in the Northeast Department—we've got
Good conditions and a decent wage.

For thirty years we have been working.
Now we've narrowed the gap.
In 1935 we started, now we're on the map.
We're from the Northeast Department—and today
We're thirty years of age.

We all must stand together and be ready to fight
To guarantee each citizen his God-given right!
Make sure indeed for race and creed to live in dignity.
Organizing hand in hand in solidarity.

[To the tune of "Solidarity Forever"]
When the union's inspiration through the workers' blood shall run,
There can be no power anywhere underneath the sun,
Yet what force on earth is weaker than the feeble strength of one,
But the union makes us strong!
Solidarity forever! Solidarity forever!
Solidarity forever!
For the union makes us strong.

From its founding to 1965 the chorus performed in sixty-five cities and towns and reached an estimated live audience of 150,000 people. It had raised $160,000 for hospitals, orphanages, civic clubs, and other charitable organizations. Bill Gable recalled:

> There was a time when the chorus performed for almost every community event. There wasn't a parade, a holiday, an important [political or social] event, or a union convention that went by where we didn't perform. We practiced a few nights every week and performed at least twice a month—sometimes more. Min Matheson would always say to me, "Bill, we have this or that event coming up or this important dignitary coming to town. Could you put something together for us?" Those really were our heydays.[32]

The chorus reflected a multifaceted agenda. Members were provided with opportunities to associate with one another and learn about teamwork. The chorus also exposed members and observers alike to ideological messages in support of the working class and the labor movement. In testimony to its significance, the group continues to the present day and remains central to the union's culture.

In December 1953 over five hundred individuals attended a testimonial dinner to honor Min Matheson at Wilkes-Barre's Hotel Redington. Speakers included Charles Zimmerman, the ILGWU's vice president; William Sword, chairman of the Greater Wilkes-Barre Chamber of Commerce; Abraham Glassberg, president of the Pennsylvania Dress Manufacturers' Association; Pennsylvania AFL officials, members of the local clergy, and ILGWU rank-and-file members. The occasion celebrated Min Matheson's achievement in building an extraordinary organization that had earned the community's respect and admiration. Min stated that she considered the affair as much a tribute to the ILGWU and the community as to herself, and she told the audience that, while much had been accomplished, a fair amount of work was still in progress and a great deal more was yet to be done.[33]

Needlepoint, the health care center, and the chorus anchored the ILGWU within the Wyoming Valley community. Yet the union's undertakings did not stop with these initiatives. Min and Bill also implemented a model workers' education program to expand the intellectual horizons of their members. And by the mid- to late 1950s, the union had positioned itself as a major force in the political arena and had staked out a role in the economic renewal of Pennsylvania's long-depressed anthracite region.

5

Constructing an Activist Union

The ILGWU has conducted all kinds of cultural activities. It would not
surprise me to learn that the union now stands as a finer
educational institution than Yale or Harvard.
—Heywood Broun, 1938

The ILGWU has been a constructive, stabilizing force in the industry and one
that has always helped in every community endeavor.
—Joseph Saporito, mayor of Pittston, 1957

The union became a real political power.
—Governor George M. Leader, 1995

Workers' Education

Education had been a hallmark of the ILGWU almost since its founding.
Within a few years of the union's formation in 1900 President Benjamin
Schlesinger initiated classes in English, citizenship, health, and history for
newly arriving immigrants assimilating into New York's garment industry
and the ranks of the ILGWU. Locals in Philadelphia and Baltimore started
similar programs. In 1916, delegates to the International's convention in
Philadelphia voted to establish an education department and allocated
$5,000 for the initiative. Juliet Poyntz, a history professor at Barnard Col-

lege and education director for Local 25, was named director. Assisting her was Fannia Cohn, a pioneer in workers' education.[1]

When Poyntz resigned in 1918, Cohn became director; she was later joined by the educator Mark Starr. Both subscribed to a humanistic educational philosophy that "constructively assists the workers to use their industrial and political power to improve their working conditions and secure necessary social changes. Basically, workers' education involves group study of social problems with a view to group action for their solution."[2]

Among the union's pioneering educational initiatives was an agreement with the New York City public school system granting use of its classrooms for "Unity Centers" and a Workers' University. Unity centers offered garment workers instruction in English, literature, reading, and physical recreation; the Workers' University offered artistic, cultural, and intellectual enrichment courses. By 1920 the two organizations annually enrolled several hundred ILGWU members.

In later years Starr initiated an Officers Qualification Program that enrolled union officers in courses ranging from labor history to contract negotiation to political activism. By the 1950s the ILGWU's educational work had expanded to include a New York–based year-round residential school for staff and officers—the Training Institute—which earned a reputation as one of the U.S. labor movement's leading educational institutions. The ILGWU's education endeavors, commented one observer, represented "the first systematic scheme of education [for] organized workmen in the United States."[3] Among the union's most significant efforts to enlighten the rank and file and provide them with leisure was a rural retreat called Unity House.[4]

Unity House

New York's Local 25—the ILGWU's shirtwaist makers' affiliate—established Workers' Unity House in 1919 to provide members with a secluded recreational and education facility. Located in Pennsylvania's Pocono Mountains near Bushkill, the former Forest Park Hotel and its 800-plus acres had previously had been owned by German-American industrialists. The compound featured a large lake, a main building with rooms for boarders, and residential bungalows set among the trees. In 1924 the General Executive Board assumed ownership and management of Unity House from Local 25, marking the first time in U.S. history that a labor union had established a facility of this magnitude solely for the leisure and enlightenment of its members. Over the next few decades Unity House expanded to accommodate over 1,000 overnight guests and became a central ILGWU institution. The union added a 1,200-seat theater with regularly scheduled Broadway-quality entertainment; dining, meeting, and dance halls; sports and recreational facilities; a sauna and fitness center; a 3,000-volume library and writing room; hiking trails; organized children's recreational programs and activities; a medical laboratory with physicians' services; a U.S. post office; and numerous guest rooms. According to the union, "In the Blue Ridge Mountains of Pennsylvania, high on a hill set in a dense forest which stretches for miles in every direction, stands our summer home. A large white house surrounded by cottages, on the shores of a blue lake, a mile and a half long, that sparkles in the sun or reflects the cool green of the overhanging trees—this is Unity House. It is a promise of a better day and our ability to bring on that day."[5]

Open only during the summer months, by the 1940s and 1950s Unity House had become one of the premier resorts in the Poconos. Yet it differed from others in this mountain vacation playland—located 80 miles west of Manhattan—in that Unity House catered exclusively to garment workers and their family members. During a typical season it hosted 10,000 visitors, many of whom returned year after year. Fannia Cohn and Mark Starr directed a full slate of formal and informal educational activities at Unity House. Lectures, workshops, and discussions, led by notable scholars, included "The Economic Basis of Modern Civilization," "Art Appreciation," "Social Psychology," labor history, American history, and many others. Since the ILGWU's educational philosophy called for workers to experience culture and the arts, the resort hosted entertainment and performances ranging from Broadway plays to the Yiddish Art Quartet, the Metropolitan Opera, and the New York City Ballet. Exposure to public policy issues also underpinned the union's educational agenda.[6] Public officials—including

U.S. senators and representatives, governors, and state and federal offi-cials—regularly visited the compound to give speeches, mingle with would-be supporters, and discuss policy issues. Unity House was also frequented by U.S. labor officials, including George Meany, president of the AFL-CIO, and the president of the United Auto Workers, Walter Reuther. Said Robert Hostetter, an ILGWU educator, "The union owned Unity House and we used it for an educational facility. Unity House came to be as some of the locals in New York wanted to get out of the city and conduct educa-tional classes. That's where it all started. And, traditionally, the different districts here on the East Coast started to use it for education purposes. It was also a vacation spot for rank and file."[7]

Thomas Mathews, who instructed at Unity House, explained both its history and its significance:

> Unity House opened after the First World War. At that time it was quite a novel thing, the notion that you would have a place where working people could go, because conditions in the factories were horrendous. Now some of this existed in Europe at that time. I believe that in Germany workers' education places existed, and also in England. But over here, it was something entirely new. I imagine that some people were outraged at the idea. I taught labor history up there. We ran a rank-and-file conference up there every year for one week.[8]

According to Marty Berger, an ILGWU veteran and guest at the Pocono resort, "Our members loved it. Our retirees loved it. Unity House was part of our culture, part of our history, part of our romance. Unity House was very important for all of us. You could go there and hear chamber music, hear a lecture, meet people like Eleanor Roosevelt. It was amazing for our people to have something like this."[9]

So central had Unity House become to the leisure and enlightenment of ILGWU members that the General Executive Board boasted of its ability to provide "rest, comfort, scenic beauty and intellectual atmosphere" and pointed out that "Unity House is the largest single ILGWU institution. It is a great cultural enterprise aiming to give our members not merely vacation

Min Matheson, David Dubinsky, and Wyoming Valley ILGWU members on the lawn at Unity House, Bushkill, 1956. (Courtesy of ILGWU.)

opportunities but opportunities for enrichment as well."[10] David Dubinsky, trumpheting its significance, required that ILGWU convention reports regularly feature—and boast about—Unity House. And when Dubinsky commissioned the making of a full-length motion picture titled *With These Hands* to depict the history of the ILGWU for its 1950 golden jubilee, he ensured that the film prominently featured a segment highlighting the resort.

Unity House not only won over garment workers but attracted the attention of the media as well. To outside observers, such as *Survey Graphic* magazine, it symbolized "the end of an era which started with the archaic European revolutionary philosophy that only by battling the bosses could the sewing machine operator and her fellow worker win the good things in life."[11] Eleanor Roosevelt visited on several occasions and was especially impressed by the accommodations for children. She commented in her syndicated column "My Day" in the *New York World Telegram* that "no matter how much of this world's goods you have, you could not put children in a more favorable environment and that is something for us as a nation to be proud of."[12]

Though many of the union visitors to Unity House came from New York and Philadelphia, various ILGWU districts and locals took advantage of the opportunities available at the resort. By the early 1950s, Wyoming Valley District members frequented Unity House as part of the larger educational agenda implemented by the Mathesons.

Education for Wyoming Valley ILGWU Members

Min and Bill Matheson provided learning opportunities to their membership that were both pragmatic and quixotic. Workers learned about community-based programs to which they could contribute and from which they could benefit. They were educated on theories regarding the industrial-capitalist order and the history and role of the labor movement. In addition, education was intended to inspire members' political and social participation.

One of the district's first educational endeavors was the Union Counselor Program, part of the Wyoming Valley's Community Chest initiative. A precursor of the modern-day United Way, the Community Chest served as an umbrella organization to raise and distribute funds for local human service and charity agencies. The Union Counselor Program educated union members who volunteered to assist and counsel rank-and-filers in need of appropriate services. College professors, union officials, business leaders, and community agency staff were among the instructors. In addition to positioning the union within the larger community, the program enhanced garment workers' financial contributions to the Community Chest. The first class of union counselors graduated in 1948. In 1953 the ILGWU raised a record $30,000 for the Community Chest. By 1961, more than 500 district members had completed the program and the union continued to raise substantial annual contributions.[13] Min made it clear that the ILGWU participated for the betterment of the community:

> A fairly large group of our members have attended the Community Chest Counselor classes and saw at first hand the work that is being carried on by the various agencies. They have visited orphanages, hospitals, Boy Scout meetings, playgrounds and the Crippled Children's Association. They have investigated the work of the visiting nurses, the family counselors, all the various Community agencies that are striving to make this valley a better place to live in. They know what the Community Chest is. They know where their dollars go and that is why they are all enthusiastic boosters of the Chest.[14]

Another initiative was the Officers' Qualification Program, first implemented in New York by Mark Starr. Bill Matheson played an instrumental role in planning and executing a program he described this way:

> The proposed "Officers' Qualification Class" is for union officers and others who aspire for office. It consists of 6 sessions:
>
> 1) The first session would deal with the start of our union, its development and expansion and the organizational changes which growth made necessary. This session would also deal with our National Convention and the General Executive Board.
>
> 2) The second session would deal with the functioning of our National Office and its various departments.
>
> 3) The third session would take up development of our own District and the Northeast Department.
>
> 4) The fourth session would deal with the New York market and our relation to the Locals and the Joint Boards.
>
> 5) The fifth session would deal with the various types of Union Contracts and how they are negotiated.
>
> 6) The sixth and final session would deal with our own communities and the role of our union in all community activity.
>
> I propose to conduct the class myself, but would require the assistance of the District Manager on some matters, such as the

rise and development of the District and its role in the Community. The presence of business agents at the class, would of course be helpful.[15]

The Officers' Qualification Program commenced within the district in early 1954. A class of twelve members graduated that spring. In 1955 locals at Pittston, Nanticoke, and Wilkes-Barre implemented the series. In Pittston, for example, twenty-one officers graduated in the spring of 1956. Fourteen more completed the requirements in early 1957. By the late 1950s, the district and its locals routinely enrolled numerous union members in fall and spring basic and refresher courses.[16]

In the mid-1950s the union established a curriculum at Wilkes College, in downtown Wilkes-Barre, consisting of credit-earning courses in the humanities and social sciences for rank-and-file members who otherwise had little opportunity to attend college. Economics, history, and labor studies were among the subjects offered by Wilkes's faculty. Min was particularly pleased with this initiative: "I worked with Dr. Farley, who was president of Wilkes College. With his help, and Dr. Rosenberg, who was a teacher of economics, we instituted a series of courses and our people went to college. And they were very proud, our women. You know, we graduated quite a few of them. They were giving them a pretty rounded view of history and so on."[17]

College-level curricula were coupled with other educational initiatives to promote an activist agenda. In one example, a voting rights class was initiated in response to suffrage inequities in towns like Pittston. Min made this point in announcing the class: "The first session of our voting class will be held at our Pittston office at 8:00 P.M, Monday, June sixth. This class will not discuss who you should vote for but will deal with the rights of the voter—particularly the woman voter—to vote without 'assistance' or interference by anybody."[18]

The Political Education Program, or PEP, launched in 1954, expanded the district's activities in linking education with political participation. Partisan in orientation, PEP encouraged members to participate in discussions on national, state, and local issues. It advocated specific policy positions, endorsed candidates, and encouraged workers to register and vote:

Wilkes-Barre's brand-new PEP Club is showing plenty of old-time pep in sparking political action. Only a few months old, PEP numbers nearly 100 active members and is fast becoming a political force in the area.

District Manager Min Matheson states that, "The disastrous effects of the present [Eisenhower] Administration's do-nothing policies are sharply reflected in this critically depressed area where workers have been hit hard by plant closings and removals. People want and need a change here for basic bread- and-butter reasons. The organized energies of PEP are geared up to help them."

Currently the club is concentrating its efforts on sending former Representative Dan Flood, who has an outstanding pro-labor voting record, back to Congress. PEP's live-wire campaign is arousing wide-spread interest and promises to develop into a dynamic community vote-getter before the home stretch is reached by November.

PEP in Wilkes-Barre holds educational meetings, checks registration of voters, distributes leaflets, places posters in factories, shops and store windows. The group recently sent a delegation of 18 members to the recent conference of the AFL Labor League for Political Education held in Harrisburg.[19]

Daniel J. Flood served as representative of the Eleventh Congressional District in 1944–46 and 1948–52. Returning Flood to Congress in 1954 became a major goal of the district. PEP and the ILGWU helped to secure Flood's successive elections to a post he maintained until 1980. Support for his policies signaled the union's maturing political voice.[20]

The ILGWU also implemented "educational institutes." These short-term courses exposed workers to political, economic, and social policy issues. Some institutes were held at Schiel's Grove, a local picnic and outing area. During the summer of 1957, for example, shop "chairladies," district officers, and delegates took part in a day-long institute featuring speakers from the ILGWU International office. Working conditions in the garment industry, state and federal legislation, and labor history were among the topics covered.[21]

The Mathesons cooperated with local institutions of higher education to

Garment workers from the anthracite region visit the United Nations, New York. (Courtesy of ILGWU.)

sponsor other institutes. During the summer of 1957, district officers attended a week-long program at the Pennsylvania State University Extension at Wilkes-Barre focusing on current labor affairs. In the early 1960s King's College in Wilkes-Barre hosted an institute that dealt with such topics as employment training, public education, the role of the state and federal governments in U.S. democracy, and labor's citizenship obligations.[22]

Recreational undertakings rounded out the district's informal educational agenda. These included modern and "farmer" dance classes complete with a graduation dance performance, writing classes, and health center–sponsored modules such as "health and charm." Also offered were classes in photography, millinery, and swimming. A factory safety and accident prevention class—including a discussion of the causes and consequences of the infamous 1911 Triangle Shirtwaist Company fire—and a first-aid program were available as well.[23]

Unity House also offered a full educational slate, in which Wyoming Valley District members participated. Building workers' awareness of political, social, and economic issues stood out as a major focus of many of the forums Clementine Lyons attended:

A performance of the ILGWU Chorus, Unity House, 1968. (Courtesy of ILGWU.)

They felt that if they could get active members together in a nice surrounding, like the Poconos, that they'd have classes up there and teach them to read, to dance, [learn] crafts. It became one of the showplaces in the Poconos. People were thrilled to come into the Poconos.

They would conduct these classes and try to develop people who would have a little bit of respect for the union. Each year they tried to do something new. They had a full movie theater right at the resort, which was nice. That was dedicated by [Governor George] Leader. It was a replica of Radio City [Music Hall] in New York. Twelve hundred seats and the lighting panel was beautiful. They put on beautiful shows up there, hired professionals or they would have us take our group [chorus] up there maybe twice a year. [There would be] speakers on politics, medical care.

Evelyn Dubrow, she would always be there. She was the [ILGWU's] lobbyist from New York, the political lobbyist. She's a good speaker. They had to have political meetings while they were there. And

Garment workers visit the U.S. Capitol, Washington, D.C., ca. 1960. (Courtesy of ILGWU.)

then, of course, they'd have one on economics, you know. Then if they had something, that was maybe when Nixon was gonna run [1960] or something like that, they'd have a little pep rally [in opposition]! They'd have that all day. You'd go down to the ballroom and listen to speeches about, you know, "Rich man, rich man," and then go down to the theater and hear the same thing all over again.

It was a beautiful place. They grew their own flowers, they had their own library. They had their own tennis courts. They had cluster cottages for families. If there wasn't a conference scheduled, they'd have other entertainment. But if they scheduled a conference, we'd even take them in buses if we had to. Because they'd bring in speakers from all over, Washington, Harrisburg.[24]

In July 1953, sixty-five district members participated in a week-long forum at the resort. Its distinctly partisan agenda included discussion of contemporary issues:

Members of the Union, spending a week at Unity House, took advantage of the occasion to brush up on current political problems as they listened to Glen Slaughter, Assistant Director of Labor's League for Political Education, on "Politics Today." Buttressing his argument with names and quotations, Mr. Slaughter hammered home the lesson that the Social Security laws, enacted by the New and Fair Deals, are now in the hands of their enemies. He urged trade unionists to join and support Labor's League (it costs only $1 per year) in its campaign to elect our friends and defeat our enemies and insure the continuation and extension of our past gains. The forum, with District Manager Min Lurye Matheson as the chair, was held on the lawn near the library and attracted members from other locals, some of whom took part in the discussion.[25]

ILGWU leaders frequently served as Unity House instructors. In the fall of 1957 members heard several key speakers:

One hundred representatives from all parts of the Northeast Department joined in a week of intensive study and discussions on such questions as The Union Shop, Ethics of Labor, Labor's Role in Politics, Civil Rights, Health and Welfare, Women's Changing Role, and Labor's History in Song and Story. Mark Starr, Director, and Marvin Rogoff, Assistant, of our National Education Department supervised the program. Fannia Cohn, the grand old pioneer of workers' education, addressed the students.[26]

A 1962 course reflected a similar approach: "The Institute covered such questions as 'Industrial Problems,' 'The Role of the Union in the Industry,' and 'Labor Legislation.' Ralph Reuter, Assistant Education Director of the International, and Sol Chaikin, Assistant Director of the Northeast Department, were in charge of the Institute. Other instructors were Gus Tyler, Education and Political Director, Alfred Gustin, Northeast Controller, and Hugh Maloney, Northeast Supervisor of Health and Welfare."[27]

Unity House often drew aspiring or prominent politicians to participate in discussions, make presentations, and socialize with garment workers.

According to the longtime ILGer Ralph Reuter, "Unity House was not only an important place for education, but it was also a place for political leaders to mix and mingle among the rank and file. Many prominent politicians came to Unity House and helped to expose garment workers to a variety of issues. The resort gave workers culture. It gave them rest and relaxation. And it gave them exposure and access to people who helped shape the policies by which they lived." [28]

Pennsylvania's governors—George Leader, David Lawrence, Milton Shapp, Robert Casey—visited on several occasions, as did mayors Richardson Dilworth, Frank Rizzo, and Wilson Goode of Philadelphia. U.S. Senators Joe Clark, Jacob Javits, and John, Robert, and Edward Kennedy spoke there, as well as Eleanor Roosevelt. Meeting such prominent public figures boosted the morale of members. It also made an impression on younger visitors, as Betty Greenberg recounts:

> We met a lot of important people at Unity House. And one of my best memories . . . I met Eleanor Roosevelt at Unity House. I was giving a speech on the lawn. I was just giving a speech to no one. I was just talking to the air. I was standing on this rock and I was going on and on about how Christopher Columbus came to this country and discovered America so that we could be free. I was going on and on. And someone said, "That's Min Matheson's daughter" to Eleanor Roosevelt. "Maybe you would like to meet her." And she came over and kissed me right here on the cheek. And I never wanted to wash it! I wanted to have it there for the rest of my life! [29]

Members were also encouraged to visit the resort simply for rest and relaxation:

> One hundred members from our District spent August 27 at Unity House. The majority of them had never been there before. They took a day off from work, rose up early and took buses to the Poconos. They had admired the beauty of Unity's 1,000 acres of forest and lawns, swam or rowed over the 100 acre lake, enjoyed two good

meals in its wonderful dining room, saw a Broadway show, danced, played ping pong and shuffleboard and met President Dubinsky and other officers of the ILGWU and AFL-CIO. Some were even interviewed by national magazines. It was a great day at Unity. Unity was at its best and it never had more appreciative guests.[30]

Through these numerous educational opportunities Wyoming Valley members gained important insight on the history of the industry, union, and larger movement of which they were a part. They experienced a level of intellectual stimulation that was a rarity on the factory floor. By their very design the programs fostered mutual association. John Justin, who organized various union educational events, remained particularly proud of the efforts in the Wyoming Valley: "The truth of the matter is that we had very fine educational programs. People who were not politically motivated became active because of what we did. They became active in the Democratic Party and in politics. We had active committees in the shops. In some cases our members ran for school boards or other things. Overall we succeeded in getting our people more alert about the issues affecting their lives."[31]

Workers' education also contributed significantly to the ILGWU's activism in politics and public policy.

Political Activism

On November 10, 1957, Mayor Joseph Saporito of Pittston issued a proclamation declaring November 16 Garment Workers' Day. The announcement read, in part: "The garment industry of the Greater Pittston area has become a major factor in our economic life, employing over 4,000 in over 40 shops. The International Ladies' Garment Workers' Union has been a constructive, stabilizing force in the industry and one that has always helped in every community endeavor."[32]

The mayor's statement recognized the many contributions of garment workers and their union. When the Wyoming Valley District was in its infancy over a decade earlier, the voting rights of Pittston-area women were in question. By the latter 1950s, however, the union was a growing political force.

Political activism had been integral to the ILGWU's agenda. Dubbed "among the most powerful political and social forces in labor," the union maintained a strong presence in local, state, and national politics.[33] Dubinsky, for example, was instrumental in forming and leading New York's Liberal Party. The union and its locals commonly endorsed candidates, advocated labor-friendly legislation, lobbied public officials, raised funds for candidates, conducted voter registration drives, and participated in political conventions. The ILGWU also maintained a strong lobbying presence in state capitals such as Harrisburg and Albany. In Washington, D.C., the union's lobbyist, Evelyn Dubrow, earned a reputation as an outspoken advocate for workers and organized labor throughout her career, which began in 1937 and continued into the 1990s. Indeed, Dubrow's work earned her the Presidential Medal of Freedom, awarded by President Clinton in 1999.

Just as union leadership realized that political activism was essential to promoting labor's agenda, the Mathesons steered their membership into the political arena. They believed that a politically engaged union could help to reform industrial capitalism and maintain a balance of power that did not relegate labor and working-class interests to second-class status. They understood that politics related to everyday matters and that linkages existed between what did and did not occur in Washington or Harrisburg and the daily lives of working people.

They secured the participation of their members in partisan processes in three ways. First, the ILGWU professed the belief that, for the most part, the Democratic Party represented the best hope for working people. The union therefore set out to encourage workers to understand that the Democratic Party represented their interests. Second, the district endorsed politicians friendly to working-class causes. Chief among them were Congressman Dan Flood and Governor George Leader. And third, the union worked through elected leaders to advocate specific legislation and policies.

Through the district's infrastructure the Mathesons set out to educate members politically. The PEP club focused on political issues. Workers' education programs were organized. The union challenged voting inequities in places like Pittston. And *Needlepoint* served as an important tool to articulate the union's views:

Dear Mrs. Matheson:

I heard you on television asking for support for the straight Democratic ticket. I am for taking part in election but I know, and I am sure you will agree, that some of the Republican candidates were better than the Democrats. How, then, could you ask for a straight Democratic vote?

Unsigned

Dear Unsigned:

I admit the question you raise has bothered me too, so let's look it over.

First, should the union be involved in politics? It is true that whatever party you support, you work in a garment shop, you are still interested in piece rates, minimum wages, hours of work and other working conditions and benefits. You are also interested in unemployment compensation, compensation in case of injury on the job, widow pensions, old-age pensions, etc. Now, some of those questions are settled with negotiations with our employer and one might think that they have nothing to do with politics. Let's have a look.

Take minimum wages in the shop. That's settled by negotiations between the Union and the employer or his association, but still, politics enters in. How? If the legal minimum is 75 cents an hour, that's what nonunion shops will pay and we couldn't get a much higher minimum from our shops if our shops are to remain in business. But if we can get the federal minimum raised to $1.25 we would have no great difficulty in getting a $1.35 minimum for you. On March 1st, the minimum goes to $1.00.

Had the Democrats lost the last Congressional elections, the minimum wage would have, undoubtedly, remained at 75 cents. Dan Flood was the Democratic candidate; Edward Bonin was the Republican candidate. We supported Flood because Flood pledged to work for a $1.25 minimum. Those who voted for Bonin were voting

to keep the 75-cent minimum. Now if you prefer the 75-cent minimum, that's your right. But please remember, one purpose of a
union is to improve wages and conditions. We were carrying out
that purpose when we were supporting Flood.

The question of which was the "best man" was not the issue. A
large part of our members have seasonal layoffs. They draw unemployment compensation. As a union, we protect the interests of our
membership even when they are laid off. How can we do that? We
supported George M. Leader and the Democratic Party in the last
State election because we wanted the unemployment compensation
raised and the interests of our members served. When Governor
Leader took office, unemployment compensation was raised, the
number of weeks extended and a provision which barred compensation to women during pregnancy was changed. In furthering the
interests of our members and improving their wages and conditions, we find that our legislative support comes in the main from
the Democratic Party. While this is true and where it is true, we shall
continue to support the candidates of that party in local, state, and
national elections.[34]

The ILGWU further argued that the Democrats were most closely aligned
to workers' interests while Republicans often tilted legislation to benefit
the wealthy:

Listen to Who's Squawking
The new chairman of the Republican National Committee and some
Republican Senators are squawking loudly about Labor's contribu-
tion to political campaigns. When the figures on campaign expen-
ditures were totaled up by a Senate Committee, the contributions
made by big business in the Republican campaign stand out like a
mighty giant of a "Goliath" against a very small, Labor "David." If
Labor's little "David," with its penny contributions, helped lay a
couple of the "Goliath" financed candidates low, it was due to the
better aim of our arguments and not to the size of Labor's penny
contributions. Republican contributions in the amount of $500 and

over, from Manhattan alone, exceed Democratic contributions from all the rest of the 48 states. The Du Pont family gave more to the Republican Party than all the contributions made to the Democratic Party from 13 southern states. In contributions from the 12 wealthiest families the Republicans led the Democrats 10 to 1. Business and professional groups, such as the Chamber of Commerce, the Bar Association and the National Association of Manufacturers, gave the Republican Party over $90 for every dollar donated to the Democratic Party. "He who pays the piper calls the tune." Apparently, the Republican big-wigs are determined that no political tune will be played that hasn't been paid for and dictated by the wealthy families and organizations of the nation.[35]

A similar theme was echoed as the 1956 primary elections drew near:

Money Talks, but People Vote
How wealthy corporations influence votes in the Senate was glaringly revealed when Senator Case told that he was given $2,500 for campaign expenses by one oil man. Reports are that Republicans are receiving ten times as much in campaign contributions as Democrats. The Republicans announced that they will have the best organized election campaign ever waged. Can money win an election? 1956 should give an answer to that question.[36]

The Republican Party was commonly associated with wealth and power, the Democratic Party with wage-earning people: "Which party is more likely to help you when elected? Just look back over the past 25 years and study the record. There are friends and foes of labor in both parties, but the party record is clear: the working men and women have made their greatest advances through the Democratic Party and that cannot be denied."[37] Democrats, therefore, had earned the endorsement of the ILGWU:

Endorsed by Labor
On the basis of their voting record and achievements and, also, on the record of achievement of the party they represent, organized

labor has endorsed the candidates of the Democratic Party for election and re-election in the state and country. Heading the group are veteran beloved Congressman Dan Flood, U.S. Senator Joseph Clark, and, for Governor, Richardson Dilworth. [We] wholeheartedly agree in endorsing the candidates of the Democratic Party.[38]

It wasn't enough to back Democrats verbally, however. Support had to translate into action. Garment workers were encouraged to exercise their voting rights:

> Who Fails to Vote?
> The well-to-do and the conservative minded citizen is usually a registered voter. But 20 percent of the working men and women either fail to register or, if they have registered, fail to vote in the elections. These stay at home workers are often responsible for the election of conservative politicians who vote against such measures as Medical Care for the Aged and Minimum Wages. The worker who doesn't vote is hurting his fellow workers in the shop.[39]

The ILGWU's rhetoric met reality in its support of Dan Flood and George Leader and in its advocacy for their worker-friendly policies.

The ILGWU and Congressman Flood

Because Dan Flood became an important ally of the ILGWU on national legislative and policy matters, union members served as his connection to the grass roots. Through their tireless support of Flood, garment workers learned at firsthand about political processes and the power of their votes. Min expanded on the mutuality of this relationship:

> Dan Flood helped us all the time. We helped him and he helped us. He's an actor. He's a performer. He became part of our performance. With his mustache and his mannerisms, he fit just perfectly into what we were doing. Just perfectly. When he became an influential

Garment workers meet with Rep. Dan Flood, Washington, D.C., ca. 1960.
(Courtesy of Kheel Center Archives, Cornell University.)

congressman we helped to elect him. We worked very hard for him
from one end of this valley to the other. And he always used to give
us credit for sending him to Congress. Well, you know, it was lot of
families and a lot of votes. He never voted wrong on any bill that
would help this area, help the unemployed, do some good. He
always voted right. I try to think of a single bill where Dan was
wrong. With all of that, he was an archpatriot. Too great a patriot in
my view, you know, because I'm critical.[40]

Flood commonly attended district events and was often a guest at union
meetings. He took part in the union's annual family picnic, which by the
early 1960s drew crowds of over 15,000. According to the organizer and
later district manager Sam Bianco, "Whenever we would have an event,
Dan was invited. He would come in late Friday night, for example, into the

airport, and he'd end up—no matter how late it was—at one of our gatherings. He never missed one of our affairs. Never. He really appreciated the help we had given him."[41] Flood proved a reliable and influential advocate on bread-and-butter issues:

> We support Dan Flood because Dan helped the garment workers. Yes. It is as simple as that. Dan worked and voted to raise the minimum wage from 40 cents to 75 cents, from 75 cents to $1, and from $1 to $1.15, and to $1.25 one year from now. Each of those raises put money in the pockets of garment workers. Each of them was won through political action. We supported Dan Flood and others like him and we continue to do so. We know that the higher minimum helped the garment workers and that's the purpose of our union— helping garment workers.[42]

The ILGWU's support of Flood was part of a growing awareness among members of the connection between congressional action and workers' daily lives:

> Dear Min,
> I would like to tell you why I like Dan Flood. Here's the reason: In 1946 Dan Flood was defeated. The Republicans won control of Congress and we got the Taft-Hartley Law. In 1948 Dan Flood was elected. The Democrats controlled Congress and we got the 75-cent minimum. In 1952 Dan Flood was defeated and we got nothing. In 1954 Dan Flood was elected and Democrats again controlled Congress and we got the dollar minimum and improvements in Social Security including reductions in retirement age for women to 62. I'd have to have my head examined if I wasn't for Dan Flood!
> An ILGer[43]

Getting out the vote for the congressman became a top priority:

> There is widespread belief here that because Dan Flood is the ideal Congressman for our district his election is assured. Elections are

never sure until the ballots are counted. That's why every member of the ILGWU should work and vote for the man who worked and voted for us. We, of the garment industry, have good cause to support Flood. In small questions and in larger issues we have had a friend in Congress and a friend who was ready, willing and able to work for us whenever we called him.[44]

Flood, in turn, reinforced the importance of citizen participation in the electoral process:

Americans, in these crucial times, can express their faith in democracy by registering and voting their conviction in the elections of 1952. Failure to register, failure to vote, is a failure in your duty to your country and its free institutions. As your Congressman and your friend, I confidently appeal to the members of the International Ladies' Garment Workers' Union in Luzerne County for a record registration and record vote this year. Make your convictions count in the best possible way, by your vote at the polls.[45]

Bill Matheson wrote simple poetry to energize the union's support for its favorite candidate (and his wife, Catherine) as elections drew near:

Dan is your man, he is tried and true.
So vote for Dan, as he voted for you.

A warm welcome to Congressman Dan and Catherine, our Valley adornin'.
With all the love that fills our hearts, we wish you top of the mornin.'

Dan's for us, we're for Dan,
Luzerne County's Congressman.[46]

The district's support of the congressman translated into promotion of his policies, particularly as they related to the economically distressed anthracite region. Flood's Area Redevelopment Bill provides an illustration.

Throughout Pennsylvania's hard coal region the anthracite-based economy had historically been subject to cyclical swings. By the end of World War II, slackening demand, inadequate capital investment, competition from alternative energy sources, and corrupt mining practices ushered in a period of steep decline. By the 1950s the near collapse of a sector that had given birth to U.S. industrialization had brought a long-term depression to the region. Mines were closing in Luzerne, Lackawanna, Schuylkill, and Carbon counties. Unemployment skyrocketed. By the mid-1950s, Wilkes-Barre remained one of two urban areas in the United States with an unemployment rate exceeding 12 percent. Neighboring Scranton was the other. Out-migration became a fact of life. The 1950 census revealed that the population of Luzerne County alone had dropped by over 41,000, or 11.4 percent, in the previous ten years, even as the population of Pennsylvania increased by 6 percent. Deindustrialization had swept the hard coal fields long before social scientists coined the term.[47]

Reversing or arresting these trends became Flood's obsession. He co-sponsored the landmark Area Redevelopment Act, which proposed to bring millions of dollars in federal aid to economically depressed Appalachian and urban areas. The bill provided tax incentives to private industry and funded job training programs and infrastructure improvements. Flood recognized that, although the garment industry was vital to his district's economic survival, industrial diversification was essential over the long haul. Without economic change, Flood feared, the area's greatest export would continue to be high school graduates.[48]

The district strongly supported Flood's legislative efforts. *Needlepoint* explained and espoused the bill's many provisions. Min detailed its positive aspects through the local media, on the factory floor, and in education programs. Members were encouraged to learn about and champion the bill among family and friends and to write the White House and Congress on its behalf.[49]

The ILGWU enrolled as a full member of the Committee of 100, an eclectic reindustrialization group consisting of local labor, business, academic, and community leaders. The group raised economic development funds and lobbied on behalf of Flood's proposal. Min testified in favor of the bill before the Senate Labor and Public Welfare Committee. Accompanied by rank-and-file representatives, she highlighted the desperate nature of the anthracite region's economic circumstance and argued that the Flood-Douglas bill, as the measure was called, provided the only real hope for a brighter future:

> When our local newspapers carried the story that I would appear before this committee, my telephone was kept constantly busy with calls from women who are deeply interested in jobs for men. You have the combined blessings of all these women. They feel that this bill of yours may be the answer to their prayers. May I say this: three out of five men unemployed in our area are skilled, experienced workers. Fifty percent of them are under 45 years of age. We have a good climate, good transportation, good water supply and all that is necessary for good, healthful living except jobs for our men. It seems to me that Government should encourage industry to locate their plants in established communities where labor is available.
>
> Let us look at the problem as it affects a worker. Here is the picture: A man works in a coal mine and is doing fairly well. He is married and has children. He takes out a loan and builds a home. His children attend a school built by him and others like him. He attends services in a church which he helped to build. He has a stake in every facility which the area offers. He is a typical American citizen, proud of his country, his town, his home. Then the mine closes down. His income stops. He registers at the Employment Service. There is no work available. He draws his unemployment compensation, but that runs out. What is he to do?

Min went on to explain the few employment alternatives that remained and asked the federal government for assistance:

> Men have gone to New York, New Jersey, other parts of Pennsylvania, in search of work, leaving their families here. In other cases

women have sought employment in garment factories, becoming the breadwinners in the homes. The greatest loss which any community can suffer is a continuation of this unhealthy and morally dangerous condition.

Surely men who have invested their all in an industry that heated the homes of the nation deserve some consideration. Our people are good people. They do what they can to help themselves and help each other. But we need aid. We need aid. I believe this bill will furnish that aid.[50]

According to one observer, "Once in a while, in the routine of covering Congressional hearings, the observer hears a witness who makes him realize that the deeper problems of America are human problems. Such a situation occurred when Mrs. Min Lurye Matheson testified before the Douglas Senate Committee on distressed areas. The guess here is that Mrs. Matheson's testimony was the kind that might propel the bill through Congress."[51]

The Flood-Douglas bill was twice amended and affirmed by Congress and twice vetoed by President Eisenhower, who had earned little respect from the ILGWU: "President Eisenhower has vetoed the Area Redevelopment Bill. This is the second time that President Eisenhower had killed hopes of bringing new industry to provide jobs for the unemployed miners of this area. One thing seems clear. Billions of dollars are O.K.'d for the development of other lands but there will be no federal aid for the displaced Anthracite miners until Eisenhower is displaced from the White House!"[52]

In 1961, President Kennedy signed the Flood-Douglas Area Redevelopment Act. Millions of dollars in aid helped the anthracite region ease its economic slide. The U.S. Commerce Department established the Area Redevelopment Administration (ARA) to administer the program. Wilkes-Barre became the agency's headquarters for the eastern United States. By 1965 the Appalachian Regional Commission assumed the duties of the ARA as Flood persuaded legislators to include all of Pennsylvania's hard coal counties within the Appalachian region. Federal funding aided, for example, plant construction and water system development for Hazleton's Valmont Industrial Park and nearby Crestwood Industrial Park, whose businesses became significant regional employers.

The Area Redevelopment Act was one of numerous Flood-initiated measures that engendered the full support and advocacy of the ILGWU. Flood served as chair of the House Subcommittee on Labor, Health, Education, and Welfare and vice chair of the Defense Appropriations Subcommittee. In these roles he influenced two-thirds of the federal budget. In classic pork-barrel style, Dan Flood saw to it that Pennsylvania's hard coal fields received its share of federal spending.

The congressman ensured that nearby Tobyhanna Army Depot became the East Coast headquarters for refurbishing military communications equipment. When federal planners drew up a design for an interstate highway to traverse the Appalachians, Flood demanded that I-81 be routed through the anthracite region. He secured the construction of a regional Veterans Administration hospital in Wilkes-Barre and procured federal financial support for an airport. Dan Flood also co-sponsored the landmark 1969 Coal Mine Health and Safety Act, which provided over $500 million in compensation to the 25,000 victims of black lung disease (or anthrasilicosis) residing in his district. By the early 1970s total federal military-industrial spending in Flood's Eleventh District amounted to an astonishing $378 million annually.

In the congressman's view, only government had the resources and the moral obligation to invest so large an amount in a depressed area. In the process of the decline of King Coal, however, a region that had once been a veritable fountain of free-enterprise capitalism had become, in large part, a ward of the state. As a result of inadequate investment, corporate and labor corruption, rancorous labor relations, and competition from other fuels, the boom of anthracite's golden age had turned to the bust of an economy that survived only through considerable government assistance.[53]

In virtually every effort put forth by Flood to revitalize the region's economy, he could count on support and lobbying by the ILGWU. According to *Needlepoint*, "Congressman Flood's long fight to get federal aid for new industrial development here was finally successful. For these and other efforts, the Congressman is to be applauded. Our congratulations and thanks go to Dan today and to President John F. Kennedy, who, like Congressman Flood, keeps his promises."[54]

The ILGWU also made significant strides in influencing state policy through their backing of George M. Leader, Pennsylvania's governor from 1955 to 1959.

The ILGWU and Governor George M. Leader

Leader's 1954 gubernatorial candidacy represented the first time that the ILGWU in the anthracite region undertook a large-scale grass-roots effort to elect a statewide office seeker. Early in Leader's campaign the union provided friendly turf for the young pro-labor Democrat in a county and region traditionally dominated by the Republican Party. Sam Bianco recalled the union's support of Leader: "When it came to George Leader, I remember I had a little white Dodge car. I remember going to the airport to pick him up. No one had endorsed him! Min Matheson and our local ILG was the first to give him any consideration of any sort. I remember going to pick him up and we went to a barbecue for him to meet some of our people. [We] decided we were going to back Leader for governor. He had a lot of obstacles to overcome."[55]

Republican dominance in voter registration numbers presented one formidable obstacle. Luzerne County Republicans outnumbered Democrats by greater than 3 to 1 (146,000 Republicans; 45,000 Democrats) in the mid-1950s. Despite the odds, the ILGWU endorsed Leader and promoted his candidacy in *Needlepoint*, on the factory floor, in union meetings, and through educational programs. Leader campaigned in area garment factories, met union members, and visited Unity House.

Though it is impossible to determine the influence of garment workers in the outcome of the election, Leader carried Luzerne County by 9,000 votes (73,512 votes vs. 64,500 for Lloyd Wood, his opponent).[56] The victory represented quite a feat, since Republican gubernatorial candidates had carried the county in virtually every election in the twentieth century. Even more astonishing was the fact that Leader was only the second Democrat elected governor in the largely blue-collar and ethnic Keystone

George M. Leader and Roy Furman, candidates for governor and lieutenant governor of Pennsylvania in 1954, meet with Min Matheson and ILGWU members in Pittston. (Pennsylvania State Archives.)

State since 1900.[57] His victory and the ILGWU's support of him were important milestones for organized labor. According to Min,

> Oh, listen, this area was Republican. Rock-solid Republican. And we, in our fashion, did a great deal to turn it from Republican to Democrat. We worked very hard to elect Governor Leader as a Democrat and that's how we broke the Republican chain of command in Pennsylvania. He gave us a lot of credit for getting [him] elected. As governor he did a lot of good, you know, for the health conditions of people and the mental hospitals. He was and still is a very honest and straightforward man.[58]

Leader realized that the ILGWU and its alignment with the Democratic Party had proved essential in breaking Republican domination in the northern reaches of the anthracite region: "When I was running [for gov-

The ILGWU Chorus campaigns for George M. Leader in his race for governor,
Wilkes-Barre, 1954. (Pennsylvania State Archives.)

ernor], the ILGWU was much more powerful than the Democratic organization. Min did a wonderful job for me. She carried a Republican county
for a Democratic candidate for governor. This was unheard of before Min
became a real political power."[59]

In his view, the ILGWU had become a major political force in the
anthracite region—a force that could sway the outcome of elections:

> My first encounter with the region was in 1952, when I ran for state
> treasurer. That's when I went into Luzerne County for the first time

Adlai Stephenson, at the microphone to the left, has the support of Min Matheson in his bid for the presidency in 1956 at this campaign rally in Public Square, Wilkes-Barre. (Courtesy of Ace Hoffman Studios.)

as a candidate. And of course later, when I ran for governor, I went back into Luzerne County.

The Democratic Party was not terribly strong in the anthracite counties. Those counties were controlled by Republicans. Luzerne County was no exception. You did not come across a lot of strong Democratic leadership. Although Luzerne County had Dr. [John] Doris as chairman [of the county Democratic Party] and he was beginning to build. It surely had been Republican for a long time. They had a Democratic mayor in Wilkes-Barre for a while. That was simply due to the fact that he was so personally popular, that he won. The Democratic organization in the anthracite area was simply not strong in those days.

David Gingold (second from left), Min Matheson, Gov. George M. Leader, and David Lawrence, mayor of Pittsburgh and gubernatorial candidate, at a campaign rally in Pittston, 1958. (Courtesy of ILGWU.)

The local governments in those anthracite counties, in most cases, were controlled by the Republicans. The ILGWU was the only Democratic force in the area. The real strength in the area that you felt when you came in was the ILGWU. There is no doubt about the fact that Min Matheson had mobilized the ILGWU as a political force that had to be reckoned with, as evidenced by the fact that they elected a Democratic congressman. Their presence was being felt.[60]

Although electoral honesty was not always common in anthracite counties, it was clear to Leader that the ILGWU's political clout and the role of organized labor were keys to his electoral majority:

There's no doubt about the fact that at that time labor unions were a lot more powerful in the elections than they are now. I had the support of the AFL-CIO, the Mineworkers, the ILGWU. All the major unions supported me with possibly one or two exceptions that I can't recall. So we had strong union support. And I carried all of the anthracite counties except Schuylkill. I probably would have carried Schuylkill but there's a good chance I got counted out in Schuylkill because, you know, they didn't count the votes too carefully in Schuylkill in those days. They just made sure that they came out with a Republican majority! In Schuylkill County you could never be sure if you won or lost by the count because you very seldom got an honest count.[61]

The ILGWU not only helped Leader win a plurality of votes in the county, but in his view, it deserved credit for Dan Flood's string of successful elections:

Dan was very much a product of the ILGWU. I don't think that, without their support, Dan could have been congressman. He was very close to Min Matheson and the ILGWU and those folks really turned out for him. I remember meeting Dan in 1952, in my campaign for state treasurer. They had a picnic and it was supported by Min and her "girls," as she would put it. It was at Sans Souci [Amusement] Park. I certainly remember meeting Dan [at the picnic]. I remember going to one of the ILGWU dinners where the president of the ILGWU was there. I attended the dinner, and of course Dan Flood was there. I'll never forget.

Dan would get up as soon as the invocation had been pronounced and start moving among the tables. Now, it was in a hall that was fairly crowded. Tables were in rows. Dan would just go up one side and down the other. He spent the entire period while the meal was being served going up and down those rows shaking hands and kidding. I watched him for a while because I admired his technique.

And those girls just loved Dan. Dan was really the product of the ILGWU. He was their sweetheart. He was their darling. He was their friend. He was the person in Washington who would speak up for them and they knew it. Min taught them that.[62]

Leader also recognized that the ILGWU served as a catalyst through which garment workers connected the politics of Washington and Harrisburg to their everyday lives:

> The ILG [was located in other parts of the state] but they weren't a political force like they were up there [in the Wyoming Valley]. Except for Min, they wouldn't have had that kind of political sophistication. They wouldn't have been informed about what takes place in Washington or how what takes place in Washington affects their lives. She had educated them. Min wasn't just a labor organizer. She was a political organizer and educator. She realized that political forces could have as much bearing on the conditions under which her girls would work as business forces. She was, without a doubt, a great force, a courageous woman. She was a great leader, a great organizer. She had great personal loyalty from her people.[63]

George Leader's election interrupted sixteen years of Republican leadership in the state's top office. Not since George Earle's term (1935–39) had labor enjoyed a gubernatorial ally. Leader's labor-friendly policies included enforcement of minimum wage and workplace safety laws, ensuring the solvency of the state's unemployment insurance fund, implementation of a Fair Employment Practices Council to police employment discrimination, expansion of unemployment insurance benefits, and job creation in economically depressed areas. Because the Commonwealth faced bankruptcy when Leader took office, he proposed a graduated income tax to garner revenue from higher incomes and investments while reducing taxation on wage earners. Though the proposal quickly gained support from labor, it resulted in one of the fiercest battles between a governor and the General Assembly in state history. When it failed, Leader was forced to expand a previously adopted sales tax to cure revenue shortfalls.[64]

From the outset of his governorship the ILGWU consistently supported Leader and his policies: "Responsible, Clear-Headed, Understanding. That's Governor Leader! Faced with a tremendous deficit piled up by the irresponsible actions of the Republicans, Governor Leader proposes a fair system of taxation; faced with depressed areas where men cannot find work, Governor Leader proposes an industrial expansion program. That's clear-headed thinking!"[65]

According to the ILGWU, Leader's election offered hope as well as a fresh approach to routine Pennsylvania politics: "We have a new Governor and a good Governor. He was the choice of the people. Supporting the Governor and his program is the first task of the Democratic Party today. That requires a united party and unity cannot be achieved in a party of the people by 'bosses rule.'"[66]

With George Leader in the governor's office, the ILGWU realized a rare opportunity to influence state policy. Economic development and job creation were at the top of the union's and the governor's agenda. To Leader, the anthracite region represented the epitome of boom-to-bust capitalism, economic decay, out-migration, and community deterioration, largely because of its reliance on a single industry:

> In the anthracite region of Pennsylvania, the issue has been the economy ever since I can remember. There is nothing else to talk about or anything nearly as relevant to those people's needs as the economy. What happened was, after World War II the young people had just left that area up there in droves. It lost a good deal of that generation, at least the male population. Now some of them came back, but a lot of them left.
>
> Some went to Washington, others to Harrisburg [for] the opportunities. The federal government was expanding during those years and they were able to move into a lot of those jobs and they were good jobs. They gave them the kind of security they never could have had in the coal region. Those departments down there were just full of people from the coal region.
>
> There is something about the exploitation [of extractive] industries that somehow, "exploitation" seems to be the only word that applies. They don't seem to care about the hospitals or the churches or the community buildings or even the infrastructure unless it directly affects them. That's what it was all about. They just never did anything to help the community. They just got in and they took

their money and they did absolutely as little as they could to protect the workers from dust, from cave-ins, from anything. They just did the minimum. Get in. Get out. Get their money and get out![67]

Like Dan Flood, the governor acknowledged the importance of garment manufacturing to the Pennsylvania economy, especially its anthracite region. However, he also appreciated the need for economic diversification. To provide state support for economic growth and diversification, he introduced the Pennsylvania Industrial Aid Bill in 1956. The measure's centerpiece was the creation of the Pennsylvania Industrial Development Authority, or PIDA, to provide low-interest loans to private firms that agreed to locate in economically distressed communities. The program targeted areas where unemployment exceeded 9 percent.

The ILGWU was one of the early supporters of the PIDA proposal. Leader acknowledged that the ILGWU played an important role in his industrial development policy:

There was a plan called the Scranton Plan. Wilkes-Barre was doing something similar where they were raising money locally to bring industry in. We were deciding what the state could do in that regard. We held a series of public hearings across the state. The first one was in Wilkes-Barre. We held one in Wilkes-Barre, one in Altoona, and one somewhere else. I remember very well the first one there, it was tremendously supported both by the people who were doing industrial development and by the unions, and ILGWU, who wanted to see more jobs. People from Scranton and Wilkes-Barre and the ILGWU came in and testified as to what the state should do to attract aboveground industry.

Frankly, out of those meetings came the Pennsylvania Industrial Development Authority. That area had a great influence on my industrial development policy. They were more advanced than any other areas of the state. They had the best ideas. And so I learned from them and we did PIDA, in part, because of the thinking, experience, and knowledge of people like the ILGWU. There is no doubt in my mind about some of the best thinking up there.[68]

Accompanied by rank-and-file members, Min testified before a state legislative committee on the need for economic diversification soon after
Leader introduced the bill. The union dispatched members to state legislative offices in Harrisburg to stimulate support when its passage was
stalled in the Republican-controlled Senate. PIDA became a topic of discussion in education programs. At the invitation of the ILGWU and Wilkes-
Barre radio station WILK, Governor Leader and Min participated in a
radio call-in program to explain the proposal.[69] Once again the union made
it clear that the interests of working people were best met by progressive-
minded Democrats like Leader:

> Governor Leader is quoted as saying that the new measure will pro-
> vide for loan assistance to community industrial development
> groups to construct and lease plants as a means of increasing
> employment in our critical economic areas. A 20 million dollar pro-
> gram passed by the Democratic controlled House is currently tied
> up in a Republican dominated Senate committee. You can help by
> electing representatives and senators to Harrisburg who will sup-
> port the Governor's progressive program. Are you registered? Do
> you vote? Do you cast your vote for the candidates who will vote
> for your needs in the House and Senate in Harrisburg? The Leader
> Administration introduced a bill authorizing aid to distressed areas
> in bringing in new industries. We consider that a step forward for
> this area.[70]

Partisan politics delayed prompt action on the bill, but eventually it was
signed into law. Pennsylvania had set a precedent by sanctioning a state
economic development policy that provided direct assistance to distressed
communities.

PIDA, the federal Area Redevelopment Act, and local economic diversification efforts yielded impressive results. Between 1956 and 1978 PIDA
approved more than 1,300 loans totaling $448 million. A large share of the
assistance went to the three main anthracite counties of Luzerne, Lackawanna, and Schuylkill. By the 1960s about fifty manufacturing firms per
year either expanded existing operations or relocated to the anthracite

The speaker's table at the Wyoming Valley District's twentieth anniversary celebration, Wilkes-Barre, 1957. Left to right, seated: David Dubinsky, Gov. George Leader, Min Matheson, Bill Matheson, Mayor Luther Kniffen, Helen Golightly; standing: Monsignor Costello, David Gingold, Rep. Dan Flood, Dr. Albert Feinberg, Rev. Jule Ayers, Julius Hochman. (Courtesy of Ace Hoffman Studios.)

region. About one-third of them were garment factories; 20 percent were in the metal trades; food and beverage establishments accounted for another 6 percent. In 1960, PIDA approved nine loans for plant expansions or relocations in the Wilkes-Barre area alone. By 1967 the immediate area gained over 13,000 new jobs for an annual payroll of over $40 million. The Crestwood Industrial Park at Mountaintop, a few miles south of Wilkes-Barre, lured fifteen new employers—including a $2 million Radio Corporation of America plant—employing over 5,000 workers.[71]

The needle trades continued to anchor the regional economy, reaching a

The chorus (far right) helps the Wyoming Valley District celebrate its twentieth anniversary in the King's College gymnasium, 1957. (Courtesy of Ace Hoffman Studios.)

peak employment of over 17,000 workers in Luzerne County and 10,500 in Lackawanna County by 1968. Moreover, combined local, state, and federal economic development efforts now enabled men and women to secure manufacturing jobs in other industries as well. In the waning years of the 1960s, as employment related to the apparel industry in Pennsylvania approached 180,000, the unemployment rate in Luzerne County hovered between 4 and 5 percent. Neighboring Lackawanna County saw its unemployment rate dip to about 5 percent. Throughout the entire anthracite region, over 80,000 new jobs were created between 1960 and 1980. The labor force grew from 180,000 in 1960 to 247,000 a decade later.[72] One regional historian noted that "in a very real sense, the community lived for twenty years—into the 1980s—on the fruits of what was accomplished in

the fifties and sixties."[73] The ILGWU, through its grass-roots political involvement, had become an important ally in the effort to re-create an economy
once dependent on a single industry.

The ILGWU continued to nurture its relationship with Governor Leader.
When the union opened a $750,000 theater at Unity House in 1956, David
Dubinsky invited Leader to assist him and AFL-CIO president George
Meany in the ribbon cutting.[74] The ILGWU endorsed and lobbied for many
additional Leader reforms, including revision of the Commonwealth's
mental health and state hospital system and implementation of educational programs for children with disabilities. The union also led voter registration drives to increase enrollment in the Democratic Party in part to
support Leader's failed bid for the U.S. Senate in 1958.[75]

Shortly after leaving office in 1959, Leader expressed his gratitude for
the union's loyalty:

> Dear Min:
>
> Thank you very much for the copy of *Needlepoint.* I am especially
> grateful for the nice things you had to say about the accomplish-
> ments of the last four years. I like to believe that governmental suc-
> cess and political success need not necessarily go hand in hand.
>
> The observations you make with regard to what was done in
> terms of helping the mentally ill, the handicapped child, and
> advances for working people, have been a source of deep satisfac-
> tion to me. That, plus the friendship of fine humanitarian people
> like you and the ILGWU, is as much as anyone could ask for.[76]

George Leader departed the state political arena in 1959. Yet the vote for
him in 1954 signaled the beginning of a significant realignment in voter
registration patterns and party dominance in a large part of the anthracite
region. Working-class and ethnic voters in several industrialized areas of
Pennsylvania had altered their allegiance from Republican to Democrat
during the New Deal, but the anthracite region did not experience this shift
until the 1950s. The trend continued over the next several years, when
Democrats became the majority in the most populous anthracite counties,
Lackawanna and Luzerne. By the early 1980s, the *Wilkes-Barre Times-Leader*

Min Matheson addresses supporters of John F. Kennedy at a rally in Public Square, Wilkes-Barre, Oct. 28, 1960. (Courtesy of Stephen N. Lukasik, Lukasik Studio.)

reported, the area's "large, blue-collar, ethnic, union-oriented populations [were] solidly Democratic." [77] The influence of Democratic politicians like Flood and Leader and the leverage of organized labor and the ILGWU were important factors in the realignment.

Meanwhile, in early 1958—midway in Governor Leader's term of office—the ILGWU had begun to prepare for its most significant strike in twenty-five years. Despite the fact that a showdown with garment manu-facturers was imminent, the union's presence in the anthracite region demonstrated that it had reached well beyond the task of organizing garment factories. As it had demonstrated in the Wyoming Valley, the ILGWU had become an activist organization concerned not only with the well-being of its members but with the welfare of communities and the equity of public policies in which working people had a stake. Indeed, observed David Dubinsky, in Pennsylvania "coal miners' garment worker wives are the latter day pioneers of the ILGWU. These women are today's counterpart of the 'old timers' [who] established the union on a permanent basis."[78]

6

Our Demands Must Be Met

The 1958 General Dress Strike

New York Mayor [Robert] Wagner called on union and employer
negotiators to . . . plea for a settlement. He warned that a [work] stoppage
would bring incalculable losses to the parties and communities.
—*New York Times*, February 28, 1958

[In 1958] many Pennsylvania dress employers, still dreaming of the old
free-for-all days, moved into a posture of open defiance.
—*Thirty-five Northeast: A Short History of the Northeast Department of the* ILGWU (1970)

Setting the Stage

Not since 1933 had the ILGWU engaged in a general industry-wide strike.
That year's work stoppage resulted in wage improvements—in some cases
as much as 100 percent—and organizing gains that added 25,000 workers
to the union's roster in New York and out-of-town locations. Hours were
reduced as well, in some cases from 49 per week to 40 or less. Ground-
breaking provisions in the 1933 agreement for the cloak and suit sector of
the industry—extended to the dress industry in 1936—held jobbers responsible for wages and piece rates paid to workers in shops with which they
contracted. The agreement also restricted jobbers' ability to change con-

tractors. These compacts enabled the ILGWU to negotiate wages and piece rates with jobbers, to enforce the rates in contract shops, and, as a result, to limit the ability of jobbers to play one contractor against another to lower wages. Though these gains were meaningful, they were not foolproof and did not stop contractors from running away from union organizers.

By the early 1950s the ILGWU's New York Dress Joint Board and five major dress industry trade associations—each of which represented jobbers and contractors in northeastern and mid-Atlantic states—had agreed to contracts that included wage and piece rate provisions, holiday pay for hourly workers, and employer contributions to the union's health, welfare, and vacation funds. The ILGWU's contract with the Popular Priced Dress Manufacturers Group, Inc. (employers who produced lower-priced dresses), was effective from March 1, 1955, to January 31, 1958, and required that jobbers use only union contractors. It likewise mandated that all jobbers register their contractors before a joint industry-union administrative board. Jobbers could register new contractors only if their inside shops or existing contractors could not handle additional work. And a jobber could not dismiss a contractor except for poor work or tardy delivery of finished goods. Folded into this agreement were many of the contractors that had been unionized by the ILGWU in Pennsylvania's anthracite region—including those in the Wyoming Valley—most of whom were members of the Pennsylvania Garment Manufacturers Association (PGMA).

The twenty-five-year industry-union peace grew tenuous when the General Executive Board met in November 1957. The board authorized the New York Dress Joint Board to call an industry-wide strike as a last resort if its efforts to negotiate a new contract failed. Six months before the GEB's authorization, at Min Matheson's invitation, the ILGWU's first vice president, Luigi Antonini, addressed Wyoming Valley dress manufacturers at a meeting held at Pittston's Lithuanian Hall.[1] Antonini—an eloquent orator and head of the 40,000-member ILGWU Local 89 in New York (the Italian Dressmakers' Union)—outlined the contractual issues as the union saw them. He informed PGMA members that they were responsible for abiding by the same wage and workplace standards as were contractors in other locales. Likewise, he said, Pennsylvania employers would be subject to the union's drive to increase wages and implement the 35-hour workweek as part of a new industry-wide agreement that the New York Dress Joint Board would negotiate with jobber and contractor associations. The jobber associations included the National Dress Manufacturers, Affiliated Dress Manufacturers, and Popular Priced Dress Manufacturers. The two contractor associations were the United Better Dress Manufacturers and United Popular Dress Manufacturers. Antonini set the stage for the battle that was to follow with contractors in the anthracite region.

The contract that had existed with the five employer associations affected over 105,000 dressmakers in New York, New Jersey, Connecticut, Massachusetts, Rhode Island, Delaware, Maryland, and Pennsylvania. It was set to expire on January 31, 1958. The ILGWU had let its demands be known: a 15 percent wage increase (there had been no general wage increase since 1953), a 35-hour week with overtime pay, holiday pay for piece workers as well as those on hourly rates, and various enhancements to contract enforcement procedures, including closer scrutiny of employers' contributions to the union's health and welfare fund.[2]

What troubled ILGWU leaders was that, despite the success of Min Matheson and other union leaders when it came to building a union in Pennsylvania, the average hourly wage of dressmakers throughout the anthracite region was consistently lower than wages in the New York market. In the Wyoming Valley, the heart of the state's dress industry, the U.S. Department of Labor reported average hourly earnings for sewing machine operators of $1.14 in 1955 as compared to $2.04 in metropolitan New York. Dress pressers earned $3.41 in New York, $1.41 in the Wyoming Valley.[3] Such wage disparities provided an incentive for New York jobbers to send more production to lower-wage Pennsylvania section work contractors. Indeed, the New York Dress Joint Board was keenly aware of the impact of low-wage competition; its membership declined by 10,000 to 42,454 from 1946 to 1956, and many of those jobs had been siphoned off to Pennsylvania's anthracite region. In addition to wage increases, the Dress Joint Board sought uniform piece rates agreed upon at the jobber's New York location to be applied to the contract shop that actually produced the garment, regardless of its location.[4]

Moreover, it was rumored that members of the PGMA—an organization that included more than ninety contractors located almost exclusively in the anthracite region—would balk at a New York–negotiated agreement that might include provisions guaranteeing uniform wages and hours. The PGMA, on the contrary, was interested in a separate agreement with Pennsylvania ILGWU districts in which its members had factories. The PGMA's intentions became clear when, in late October 1957, employer members closed their shops and locked out their workers for three days. The event culminated in an announcement that the PGMA would resign its affiliation with the jobber-dominated Popular Priced Dress Manufacturers Association, which had bound PGMA members to agreements negotiated with the New York Dress Joint Board. After the PGMA's announcement, seventy New York jobbers—all of whom supplied work to PGMA shops—made it known that they, too, would resign their association affiliation and not be bound by any contract negotiated.[5]

Julius Hochman, vice president and general manager of the Dress Joint Board, joined Northeast Department director, David Gingold, in criticizing PGMA members for their stance: "This group of firms, made up of employers that do contract work for New York, has raised the cry that it wants a separate agreement with the ILGWU. This is despite the fact that for the past 15 years its members have lived and prospered under an industry-wide set of collective contracts that include dress shops in New York [City], New Jersey, parts of New England, and Upstate New York."[6] The union was convinced that a separate contract with the PGMA would be less stringent and permit Pennsylvania employers to evade the obligations to which all other jobbers and contractors would be held.

Union delegates met in Bethlehem, Pennsylvania, on February 26 to discuss strike mobilization. Delegates from 300 factories in the state agreed to form strike committees and set up strike machinery. At the meeting, Director Gingold commented that while the ILGWU and the dress industry trade associations were working to resolve their differences, the Pennsylvania contractors remained obstinate and outside the negotiations.[7]

In the meantime, the contract's expiration deadline was extended to February 28 to allow David Dubinsky, Julius Hochman, and other ILGWU negotiators to meet with industry trade association representatives at New York's Park Sheraton Hotel. As the February 28 deadline approached, Dubinsky announced his intention to continue around-the-clock negotiations to reach a settlement.[8] He was willing, he said, to compromise on the ILGWU's demand for a 15 percent across-the-board pay increase but would stand firm on all other demands. As Friday, February 28, drew near, the union reiterated its strike threat and indicated that a work stoppage would be delayed only so long as the contractors' consortium was willing to continue discussions. Dubinsky told the press that the conferees had decided to "explore the possibility of compromise."[9] While the ILGWU was willing to make concessions, however, employers' opposition made a settlement nearly impossible as the weekend drew to a close. An offer by New York Mayor Robert Wagner to mediate the matter was turned down by the parties.[10]

On Saturday, March 1, 1958, a final meeting between labor leaders and employers' representatives took place in New York. The assembly lasted less than one hour. The union rejected the industry's offer of a 5 percent wage increase. Dubinsky was equally unsatisfied with the industry's intention to recognize 37 1/2 hours as an official workweek and to pay overtime for additional hours. Contract enforcement provisions, designed to secure wage parity between metropolitan and nonmetropolitan markets, were also not up to the ILGWU's expectations.[11] As a result, the union announced that a strike would commence the following week.

Throughout the anthracite region estimates were that 11,000 ILGWU members in 250 factories would be affected by the strike, well over half of them employed in 140-plus factories in the Wyoming Valley. Not only would dressmakers be out of work but entire factory staffs, truckers, and suppliers were to feel the impact—in excess of 25,000 people in the Keystone State. And dress retailers, wholesalers, and customers would be affected by depleted inventories if the strike were prolonged.[12]

The public weighed in on the impending strike. In an anonymous letter to the editor, a concerned citizen wrote: "There is no reason to believe that anyone connected with this controversy actually wants a cessation of operations. Nor should a shutdown be regarded as inevitable or the only solution to the problems. There is always a solution to every conflict if it can be found in time to prevent a shutdown. There has to be a solution if the industry is to survive, so why not now before everybody is hurt?"[13]

The *Wilkes-Barre Times-Leader* opined:

It is hoped that cool heads will prevail in the current dispute that threatens to erupt into a strike. The entire anthracite region . . . can ill afford a shutdown of the ladies' dress industry, particularly at this time of recession. A strike will mean only hardship of the workers and their dependents and an unrecoverable loss for their employers. It will be a blow to the public in this region because of the adverse effect the payroll loss will have on our economy. Nobody stands to win by halting work now or at any time. The ILGWU has enlightened leadership. The association of employers is equally advanced in its collective thinking. The issue which has so far blocked an agreement may not be as insurmountable as it seems if the parties concerned sit down together with the will to do the fair thing. Continued negotiations are in order until the impasse is broken. Nothing else will reflect credit on either side. The public awaits a demonstration of responsibility by both parties to the dispute.[14]

The Wyoming Valley's Labor-Management-Citizens Committee took its own initiative to prevent the walkout.[15] Committee representatives met with members of the PGMA and the ILGWU in an effort to head off the shutdown in local plants and shops. Min Matheson told the group of citizen, business, and labor leaders that even though industry-wide negotiations were continuing in New York, the possibility of agreement had all but disintegrated. Moreover, she felt, negotiations with the PGMA were at an impasse. Indeed, she had already ordered members to walk off the job on Wednesday, March 5.[16]

In preparation for the first dress industry strike in twenty-five years, the ILGWU informed its members employed in PGMA shops that "all members of the ILGWU employed in dress firms and member shops of the Pennsylvania Garment Manufacturers Association are not to report to work Wednesday morning [March 5]. They are instructed to report to their assigned strike headquarters in the respective areas in which they work. They are to check with their chairmen for affirmation." For New York workers, the strike was set to commence with the beginning of the first shift. Employees were instructed to report to their jobs at their usual starting

David Dubinsky delivers a fiery speech explaining the union's demands in the general dress strike, 1958. (Courtesy of Kheel Center Archives, Cornell University.)

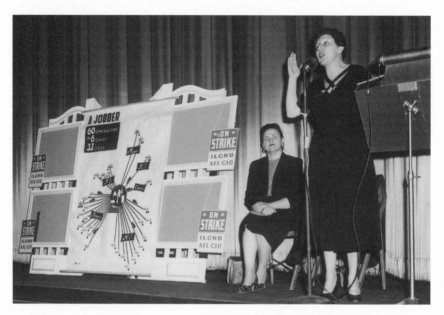

Min Matheson addresses Wyoming Valley strikers in preparation for the 1958 general dress strike. (Courtesy of Stephen N. Lukasik, Lukasik Studio.)

times, declare that they were on strike, and proceed to Madison Square Garden for a mass meeting.[17]

Essentially, two strikes began on March 5: one against seventy New York–based dress jobbers, the entire PGMA, and nonunion employers; the other against employers affiliated with the five major dress industry associations. For the first time in a quarter of a century, the core of the nation's $6.3 billion dressmaking industry was brought to a standstill.[18]

Strike

One hundred and five thousand workers walked off the job with the day's first shift.[19] For the duration of the stoppage, picketers were placed at every closed shop from 8:00 A.M. to 6:00 P.M., and patrols were out each night from 6:00 P.M. through the morning to prevent any shop from operating. Because spring and the Easter holiday were approaching, the demand for seasonal clothing provided an incentive for a quick settlement. Speculation was widespread that an agreement would be reached by the coming weekend. New York City Labor Commissioner Harold Felix said that "the hope [is that] there will be an acceleration on both sides toward joint negotiations."[20] *Women's Wear Daily* informed its readers: "The chances of reaching an agreement before next week were held to be good, according to an informed source. It was said that the difficulties pertaining to the union's demands for stronger enforcement will be cleared up, and management will boost its wage increase offer."[21]

Meanwhile in Pennsylvania, Min Matheson reported that "the walkout is 100% effective at all the struck plants," implying that no production had taken place at any of the factories affected by the work stoppage; "we're here to stop every shop in Wyoming Valley until a new contract is signed." Support for the strikers came from many quarters, including local clergy. The Reverend Edward F. Singer gave the strikers words of support: "If I may speak for the community, I would say that the people of Pennsylvania sympathize with the aspirations of the workers in the dress industry and sincerely hope that conditions of labor and rates of pay may be arrived at which will mean the progressive realization of the 'good life' for all who labor. All of us involved with labor—organized and unorganized—do not want to see labor held down to any ancient serfdom."[22]

Pleased with strikers' response in the anthracite region, Julius Hochman complimented Harry Schindler, Scranton district manager:

> Thanks for your heartening report that the dressmakers in your area have responded with a unity and solidarity to the general strike call. Over 100,000 workers have joined the walkout and brought operations in the industry to a standstill. The Madison Square Garden meeting in New York was a tremendously successful demonstration of the workers and the strength of our organization. Together we shall achieve victory and write another glorious chapter in the history of out great union.[23]

Problems developed in the anthracite region, however. The owners of three shops in Pittston attempted to resume operations. Angelo Sciandra, Dominick Alaimo (both alleged members of organized crime), and William Alba, owners and managers of Ann Lee Frocks, Jane Hogan Apparel, and a plant in nearby Port Griffith, respectively, opened their doors on the second day of the strike and several women showed up for work. Word reached the ILGWU and picketers promptly arrived at each factory to pressure workers to shut off their sewing machines and not "scab." The actions proved successful. Pittston also laid claim to the first violence of the strike. Approximately twenty-five strikers entered a nonunion dress factory, threatened the employees and owner, and reportedly caused $15,000 in damage to equipment and materials. Later during the first week of the strike, workers at fifty nonunion dress factories—over half located in Pennsylvania—joined in the walkout.[24]

With the assistance of Frank Brown, a mediator appointed by the U.S. Department of Labor, negotiations between employers' associations and ILGWU representatives resumed in New York City on March 6. When Brown reported little progress, Mayor Wagner called a meeting of the parties on March 7 at City Hall. Wagner asked Governor Herbert H. Lehman and Harry Uviller, chairman of the New York State Mediation Board, to direct the talks. Concerned that a drawn-out strike would imperil one of New York's major industries, Wagner was sending a clear signal to both sides about the sincerity of his interest in a quick settlement; union leaders and employers assured him that everything possible would be done to reach an agreement.[25]

Wagner's intervention proved fruitful. A three-year pact was produced—for all but PGMA shops—on Tuesday, March 11. The new agreement included wage increases of 8 percent, a 35-hour workweek, time and a half for work in excess of 7 hours per day, additional holidays, the establishment of an employer-financed severance fund, and the requirement that all garments produced in union shops bear the union label.[26] Additional enforcement language incorporated into the agreement strengthened the union's ability to enforce piece rates settled with jobbers for apparel produced in contract shops. The new accord allowed garments to be produced and shipped in time for the spring season.[27]

In Pennsylvania, however, achieving a settlement proved far more difficult.

The Pennsylvania Holdouts

Despite the settlement, the PGMA had still made no effort to negotiate with the ILGWU. "Thus," said Dubinsky, "the strike is not over in Pennsylvania; it will continue with our money and our sacrifices."[28] It appeared that the PGMA and the ILGWU were no closer to agreement than they had been in January. In the Wyoming Valley—the heart of PGMA territory—about 1,700 of the area's 6,000 strikers had returned to work at nineteen independent plants; picketing continued unabated at all PGMA shops. In addition to picketing, strikers maintained the ILGWU's high profile by visiting patients in local hospitals and distributing gifts to sick children.[29]

The union continued its unabashed criticism of the PGMA. David Gingold told garment workers gathered in Pittston that they had the ILGWU's full support and that the union would advance strike benefits to those out of work. On behalf of the New York Dress Joint Board, Julius Hochman echoed Gingold's commitment. Gingold held out hope that a potential resolution lay in the desire of many PGMA shops to resign from the association and sign independent contracts with the union. The Mathesons likewise criticized the PGMA's stubbornness, claiming that even the name of the association was deceitful. As contractors, Min said, the association should be renamed the "Pennsylvania Contractors' Association. They are not true manufacturers." She added, "They contract work from New York jobbers and bring it to the valley for you to sew."[30]

The strike remained front-page news in much of the anthracite region in mid-March. Although dressmaking in Manhattan had returned to near-normal production levels, about 15,000 workers—from sewing machine operators to truckers, office workers, managers, and suppliers—remained out of work in northeastern Pennsylvania and some parts of upstate New York (where PGMA members also had contract factories).[31]

Prospects for a settlement looked grim as the PGMA's board and membership met in Hazleton. They announced that they would "fight to the end," and insisted that any bargaining would take place only between PGMA and representatives of Pennsylvania districts of the ILGWU. The New York Dress Joint Board, according to the PGMA, was not welcome at the bargaining table.[32] The ILGWU, on the contrary, was adamant in its resolve that any settlement with the Pennsylvania contractors had to match the agreement with the rest of the industry.

Talks between the PGMA and the ILGWU, arranged with the assistance of Wilkes-Barre's Labor-Management-Citizens Committee, took place in New York. Dubinsky, Gingold, Matheson, and other ILGWU leaders met with PGMA president Abraham Glassberg and attorneys Israel Klapper and Max Rosenn at union headquarters on Broadway near Times Square. Before the meeting PGMA officials said they would consider a contract with the ILGWU that would essentially match the provisions agreed to by the New York City dress industry. They would not, however, sign such a contract with the New York Joint Dress Board—instead, they wanted separate contracts with the union's Pennsylvania districts. Separate agreements were unacceptable to the ILGWU, as they could open the door for future deviations from an industry-wide contract. Not surprisingly, the conclave concluded with no agreement.[33]

Meanwhile the PGMA requested a hearing before the National Labor Relations Board (NLRB) "to determine what union represents its employees." Association attorneys argued that it was not possible for the employers to deal with both the ILGWU in Pennsylvania and the New York Dress Joint Board. They requested the NLRB to rule that Pennsylvania employers bargain only with the ILGWU districts in the state. Though the NLRB postponed a hearing, some workers trickled back to their sewing machines. Eighty ILGWU members in Exeter Borough returned to work in addition to 200 in Wilkes-Barre and Plymouth when their employers split with the PGMA, signed agreements with the ILGWU that included provisions similar to the industry-wide settlement, and reopened their factories.[34]

In its efforts to maintain pressure on the PGMA to honor a contract similar to the industry-wide agreement, Julius Hochman told a crowd of strikers that the union's strategy was to deal with factory owners not as a conglomeration but as many individuals: "As of today, we will settle with any

Contract negotiations between representatives of the ILGWU and the Pennsylvania Garment Manufacturers Association (PGMA). (Courtesy of ILGWU.)

contractor who wants to break away from the PGMA. If we can't settle them all in a bunch we will pick them up one by one. Why does the PGMA fight? I'll tell you why they fight. They fight for pennies—your pennies, there is no other issue. The employers have not conceded that there has to be justice, reasonableness, and that they have to treat people as human beings." [35]

While some owners signed individual agreements with the ILGWU, several began to withdraw from the PGMA and affiliate with the United Better Dress Manufacturing Association of New York, already under contract with the union.[36] By March 25 a total of twenty-seven area factories, formerly associated with the PGMA, resumed work under a new contract. By early April, the number neared forty, including seven factories in the Lehigh Valley that were part of the union's Allentown/Reading district. The April issue of the garment workers' paper *Justice* proclaimed:

About 75% of jobbers who broke away from their industry association before the start of the general strike, and who engineered a
simultaneous walkout of PGMA contractors from their parent
United Popular association, have reaffiliated themselves [with
United Popular] and want to work. Reports from Pennsylvania
indicate that about half of the original break-away contractors in
that state have accepted a means for returning to work and that
most of the others also want to get back to work.[37]

Rumors of a possible agreement were afoot in early April. David Gingold attempted, once again, to negotiate with PGMA leadership. According to the union, terms of a possible contract would require the PGMA to
rejoin the United Popular Dress Manufacturers Association, agree to a 35-
hour workweek, and raise pay scales to narrow the difference between
Pennsylvania and New York workers. The PGMA would also agree to
withdraw its request for an NLRB hearing. When pressured to agree to
such terms, however, the PGMA leadership balked.[38]

As the dispute dragged on, violence became more frequent. A picketer
at Wyoming Frocks was punched in the chest by an unknown assailant,
fainted, and was taken to the hospital. The next day, at the same shop, two
women were kicked and punched while picketing, allegedly by the two co-
owners of the company. They were not seriously harmed.[39] Min Matheson
and three others were taken by police to Pittston City Hall from a picket
line at Jenkins Sportswear in Pittston, a dress factory owned by Emanuella
Bufalino Falcone, niece of organized crime boss Russell Bufalino. According to press reports, when strikebreakers attempted to cross the picket line,
a scuffle ensued, police were called, and the ILGWU contingent were
arrested and charged with aggravated assault, battery, and attempting to
incite a riot. Pittston's chief of police released the detainees on the condition that they return for later hearings. Matheson charged that the whole
affair constituted nothing more than "intimidation of picketers by men of
questionable character hired by Russ Bufalino." She decried the "police
brutality against pickets."[40]

On another occasion, when the owners of Pittston's Lori Dress Factory
attempted to haul dresses by truck to New York, a struggle broke out

between the union picketers and the truck drivers. The picketers flung rocks through the trucks' windshields, deflated the tires with hairpins, and pulled wiring out from under the dashboard. Wary of taking physical measures against women, police officers and sheriff's deputies did not interfere, though the trucks were too damaged to be driven. When yet another call was made to the ILGWU's Pittston office to report the presence of two more trucks at a plant's loading dock, over 200 picketers, all women, again arrived at the scene. The drivers climbed down from their trucks after nearly crushing five picketers between their bumpers and the loading dock. A plant manager waited near the loading dock with pistols in his hands. Police arrived on the scene and ushered the armed men back into the factory. Meanwhile, strikers deflated the trucks' tires, poured sand into their gas tanks, damaged the electrical systems, and threw the keys into the nearby Susquehanna River. When another truck arrived on the scene to carry away a load of dresses, the panicked driver quickly drove away.[41]

Vandals also attacked the Mathesons' home with red paint bombs during the early-morning hours of April 8. Min, Bill, and their daughters had gone to bed at 1:30 A.M. and awoke at 7:30 A.M. to discover a shattered glass container near the front porch and red paint splattered across the front of their Kingston house. A Wilkes-Barre police officer simultaneously discovered that a red paint bomb had been lobbed at the union's health care center on South Washington Street.[42]

Reacting to the ongoing strike, the Pittston uprising, and the potential for further violence, Luzerne County Sheriff Joseph Mock declared a countywide state of emergency. Curiously, the sheriff's proclamation was targeted at the ILGWU with no mention of the PGMA or any factory owner. The proclamation read, in part:

> Whereas . . . Deputy Sheriffs [have] met with unlawful opposition and acts of violence in . . . Luzerne County, Pennsylvania, and
>
> Whereas . . . these acts constituted obstruction of officers in their duty and,
>
> Whereas, certain officers of the International Ladies' Garment Workers Union together with numerous members of that Union have been main and principal offenders,

Now, know ye, that I, Joseph Mock, do hereby declare that an emergency exists in the County of Luzerne, and to prevent . . . the officials of said [union] and its members from obstructing or in any way interfering with the Sheriff and his Deputies in [the] lawful service . . .

All good citizens, officers of the [ILGWU] and its members will observe the law: and others are warned that they must obey the provisions of this emergency.[43]

Several days before the sheriff took action, the NLRB commenced a hearing in Newark, New Jersey, on the issue of union representation for employees in Pennsylvania factories. The PGMA argued before the NLRB that employees in its members' factories should be permitted to cast new votes for union representation. No contract was in force with the ILGWU's New York Dress Joint Board, as it had been before. Therefore, the PGMA said, lacking official union representation, Pennsylvania workers should be permitted to choose their collective bargaining agent—namely, the Dress Joint Board or a Pennsylvania district of the union. The ILGWU retorted that such a vote was unnecessary. Workers had already validated the ILGWU, making no distinction between the New York Dress Joint Board and Pennsylvania districts. PGMA countered that the ILGWU feared such a vote only because there was reason to believe that the Dress Joint Board would lose its bargaining authority. The hearing also touched on wage issues. Irving Jackman, chairman of the board of the PGMA, agreed that Pennsylvania dress manufacturers had not compensated employees on a par with those in New York, nor did they intend to do so, regardless of any collectively bargained agreement. Jackman's view was that Pennsylvania employers were not obliged to comply with the terms of any contract that covered New York workers.[44]

The ILGWU persisted in its strategy to reach agreement with employers independent of their association with the PGMA. Displaying his dissatisfaction with the association's efforts, Gingold said that he had "had enough of the Pennsylvania Garment Manufacturers Association." The PGMA was determined to keep the owners unified; contracts with individual firms would "ultimately lead to the destruction of the dress industry in Pennsylvania," asserted its president, Abraham Glassberg. "The union's aim for individual agreements indicated the evident purposes from the very beginning to weaken and ultimately destroy the association because they know full well that individual Pennsylvania employers are too weak economically to resist the unreasonable demands of the union."[45]

According to David Gingold, however, the ILGWU apparently had no other choice: "It has become clear that it is futile for the so-called leaders of their [PGMA] association to conduct conferences when that leadership confers in bad faith and the understandings they reach are evidently vetoed again and again by powers unknown to us." Gingold was referring to apparent backpedaling by the PGMA when it reneged on an earlier pledge to abide by an agreement with the ILGWU based on the larger industry contract. "The leader of the PGMA [Glassberg] solemnly signed an agreement calling for a settlement of the strike within the framework of the industry-wide agreement in which the special problems of Pennsylvania, including a [wage] differential, were agreed upon. They [later] reneged and rejected the [agreement] which they had signed."[46]

According to the *New York Times*, concurrent with the signing of an industry-wide contract, PGMA representatives had, in fact, signed an agreement with the ILGWU that included all of the union's demands, including a wage increase. A piece-rate "discount" was included in the agreement, however, to compensate Pennsylvania employers for the costs of trucking and special equipment. Yet, according to the *Times*, "even before ratification, word circulated in [New York's] garment district that [PGMA's] 'invisible government' had vetoed the pact" and that the main figures in this so-called invisible government were the underworld figures Russell Bufalino and Thomas Luchese, owners of dress factories in the anthracite region.[47]

Despite alleged mob influence in the PGMA, the union maintained its shop-by-shop settlement strategy. On April 6 Min Matheson called thirty factory owners to a meeting and proposed that they adopt the industry-wide contract. The factory owners rejected her proposal. When the discussion grew heated, Min—pounding a table and breaking several water glasses—threatened to go to the press to "tell the public what kind of men you are" and expressed the union's concern that if work did not resume in the very near future, New York jobbers would probably take their work elsewhere, in which case, everyone would lose.[48]

On April 15 David Dubinsky spoke at rallies in Scranton and Pittston, blamed several underworld figures by name for blocking a settlement, and announced that the final day for negotiations had arrived. "If they don't settle today," he proclaimed,

> they can get out of the business and we'll find other contractors who will abide by the industry-wide agreements. They are trying to isolate you so they can use you as slaves.
>
> When things got too hot in New York they found an outlet in Pennsylvania. Today a settlement evaporated because of some underworld characters—hoodlums and their stooges. But they will not take the bread out of your mouths. Not a dress will be manufactured until an agreement is signed. We refuse to deal with underworld characters or their stooges. If there's racketeering, over my dead body will they rule this industry.[49]

Dubinsky further alleged that some contractors were intimidated into holding out against union demands. Immediately after his rousing speech to the workers, Dubinsky entered into a meeting with PGMA officials for the last time: "After two and one-half hours," the ILGWU reported, "he left the conference and announced no settlement. He returned to the conference at the request of [Pennsylvania Department of Labor and Industry] Secretary [William] Batt and soon after, exited again, this time with the definitive announcement that a complete break had occurred and that he would not again participate in any conference with PGMA representatives."[50]

The strike in Pennsylvania officially ended on April 17, when 60 contractors relinquished their ties to the PGMA and signing individual agreements with the ILGWU. More would follow, and by the end of April only 15 contractors and 11 jobbers out of the original group of 190 contractors and 70 jobbers remained on strike. Most would eventually settle.

The terms of the individual agreements generally reflected those incorporated in the industry-wide contract: a 35-hour workweek, time and a half for

The 1958 General Dress Strike

Min Matheson and David Dubinsky after the settlement of the 1958 general dress strike. (Courtesy of ILGWU.)

work in excess of 7 hours per day, additional holidays, an employer-financed severance fund, and the requirement that all garments produced in union shops throughout Pennsylvania bear the union label. On the issue of wages, the individual agreements included increases of as much as 8 percent. However, piece rates in Pennsylvania were to be subject to a "discount" of about 10 percent from those negotiated by the New York Dress Joint Board with that city's employers. The discount was intended to compensate Pennsylvania employers for costs associated with trucking and machinery.[51]

The PGMA's president, arguing that the ILGWU was attempting to "impose a stranglehold on the garment industry in Pennsylvania," concluded that the association had no choice but to relinquish its hold on affiliates and thus pave the way for individual agreements:

I had no alternative but to release the members to sign individual contracts due to the economic pressure exerted by the union. I am firmly convinced that the association could not undertake to sign a contract guaranteeing the union that members would comply with the terms of the agreement as signed in New York. The association would become a police agency to compel its members to comply with this agreement. The individual plant owners are small businessmen and are unable to withstand the economic pressure of a long strike. The test will come in about six months when the public and the employees will realize the full effect of the union's campaign.[52]

Among the remaining diehard firms were those owned or controlled by Russell Bufalino and "Three Finger" Brown (aka Tommy Luchese). Indeed, as Dubinsky asserted, the union was experiencing difficulty in settling the strike against certain Pennsylvania employers because of interference by "underworld characters."

During the general strike, the ILGWU learned that Pittston's Jenkins Sportswear was employing nonunion labor to continue to produce women's dresses for a struck New York jobber. Picket lines were set up immediately and the owners responded with violence against picketers that continued for several months.[53] Matheson wrote to Secretary Batt, head of Pennsylvania's Department of Labor and Industry: "Today, [a] group attacked our women pickets, injuring several of them and also viciously beat up a man, a passer-by, who attempted to intervene. The ILGWU considers this the beginning of Bufalino's war on our Union, and of course, we will not allow it to succeed."[54]

Nine individuals were indicted for assault against picketers, including Russell Bufalino, whose niece, Emanuella Bufalino Falcone, was principal owner of the factory. In a countermove against the ILGWU, William Bufalino, a relative of Russell, established a competing union, the Northeastern Pennsylvania Needle Workers Association. The ILGWU's dispute with Jenkins Sportswear extended through a series of legal maneuverings, charges, countercharges, and violence that extended for another eighteen months. In December 1960 the union and Jenkins settled the dispute with a contract that extended the 1958 agreement to the Pittston factory and its employees.

Min Matheson addresses strikers from the back of a pickup truck as the ILGWU strikes Jenkins Sportswear, Pittston. (Courtesy of Stephen N. Lukasik, Lukasik Studio.)

Throughout the dispute Min Matheson and Russell Bufalino remained bitterly at odds, and tension between union and employer remained apparent even after the parties reached a compact.[55]

In a similar situation, the ILGWU battled a holdout factory in Sweet Valley, north of Wilkes-Barre, owned by New York mobster Thomas Luchese. When Havric Sportswear continued to produce women's dresses with nonunion labor, the ILGWU set up twenty-four-hour pickets and rallies at the rural factory. In one instance, when plant management attempted to ship dresses to New York, strikers disabled the truck by putting sugar in the gas tank and placing nailed blocks under the wheels. In another incident strikers hijacked a truck and its cargo and hid it in a remote area until the union's conflict with the employer could be settled. Luchese reacted to the ongoing dispute by selling his interest in Havric to a local owner. The shop reopened with an ILGWU contract.[56]

Bufalino's eventual concession to the ILGWU and Luchese's sellout were at least in part influenced by increasing attention from the federal government. Concurrent with the ILGWU's disputes with both mobsters, organized crime's interests in apparel manufacturing in northeastern Pennsylvania drew the attention of the U.S. Senate Select Committee on Improper Activities in the Labor or Management Field and Robert F. Kennedy, its chief counsel. The McClellan Committee formally launched an investigation into connections between the underworld and apparel making in the region and confirmed Bufalino's interests in at least five factories as well as Luchese's interest in Havric Sportswear. Although Bufalino and Luchese presented themselves as legitimate businessmen, the committee concluded that both held considerable influence and power in the underworld.[57]

Several months after the commencement of the 1958 general dress strike, about 400 strikers were still walking the picket lines around a handful of PGMA affiliates who simply refused to consent to industry-wide contract terms.[58]

A Partial Success

The 1958 strike demonstrated the power of the nation's leading apparel workers' union to virtually shut down the entire dress industry in New York, Pennsylvania, and other northeastern states. Much was gained: a shorter workweek, wage increases, and recognition of the soon-to-be-famous ILGWU union label. In the longer term, however, wage equity between Pennsylvania and New York workers—a vitally important goal to the ILGWU—was never fully achieved. The threat remained that the availability of lower-wage workers would continue to draw jobs away from Manhattan's garment center. Indeed, the availability of lower paid workers elsewhere in the United States and overseas would continue to threaten the union's gains and ultimately its very existence.

Eleanor Roosevelt symbolically sews in the billionth ILGWU union label, 1961.
(Courtesy of Kheel Center Archives, Cornell University.)

Although wages for garment workers in Pennsylvania and the anthracite region continued to rise, the union had difficulty in enforcing centralized Dress Joint Board piece rates. Board officers blamed Pennsylvania ILGWU staff for failing to enforce the rates out of fear that employers would react by engaging in new runaway tactics to find less costly labor—a fear soon realized. Pennsylvania staff, including Min Matheson, accused local contractors of noncompliance and blamed the union's Price Settlement Department—an internal organization that worked with industry representatives to determine appropriate piece rates—for not providing timely and accurate piece rate settlement data.

Yet the wage gap between New York and Pennsylvania workers was indeed narrowed. While Pennsylvania dressmakers earned only 56 percent of the wages paid to their New York counterparts in 1955, by 1960 the gap

Table 2 Average Weekly Earnings of Pennsylvania Apparel Workers, 1956–1972

Year	Pennsylvania[a]	Scranton/Wilkes-Barre/ Hazleton MSA[b]
1956	$45.33	$43.87
1960	53.48	46.57
1964	60.70	44.36
1968	70.30	66.50
1972	93.07	83.64

[a]Pennsylvania Department of Labor and Industry, Bureau of Employment Security, *Pennsylvania Employment and Earnings* 1, no. 1 (January 1956); 6, no.1 (January 1961); 10, no. 1 (January 1965); 14, no.1 (January 1969); 18, no. 1 (January 1973). These figures cover all workers employed in all sectors of the apparel industry in Pennsylvania, including members of the ILGWU, Amalgamated Clothing and Textile Workers Union, and nonunion workers.
[b]Pennsylvania Department of Labor and Industry, Bureau of Employment Security, *Scranton Labor Market Letter* 11, no.1 (January 1956); 15, no. 1 (January 1960); 19, no. 1 (January 1964); 23, no. 1 (January 1968); 27, no. 1 (January 1972); and *Wilkes-Barre/Hazleton Labor Market Letter* 11, no. 1 (January 1956); 15, no. 2 (January 1960); 9, no. 1 (January 1964); 23, no. 1 (January 1968); 27, no. 1 (January 1972). These figures cover all workers employed in all sectors of the apparel industry in the Scranton, Wilkes-Barre, and Hazleton Metropolitan Statistical Area, including members of the ILGWU, Amalgamated Clothing and Textile Workers Union, and nonunion workers.

was reduced to 67 percent. By 1963 the differential was 82 percent and by 1968 about 89 percent.[59] (See Table 2.)

By the early 1950s—several years before the general dress strike—garment manufacturers had begun to move production to the southern United States. Two advantages lured them: the availability of workers employable for wages lower than those paid in New York, Pennsylvania, and elsewhere in the northeastern United States; and the lack of unions or, in some cases, hostility toward them. The industry's movement to the South, however, proved to be merely an interim step.

Meanwhile, in January 1963, David Dubinsky issued an announcement that startled members of what had become the union's second largest district in the Keystone State:

> Min Lurye Matheson, manager of the ILGWU's Wyoming Valley District of the Northeast Department in Wilkes-Barre, Pa., since 1946, has been named director of the garment workers' National Union Label Department.

Mrs. Matheson was chosen unanimously for the post by the Union Label Committee of the union's General Executive Board, acting on a recommendation of Pres. David Dubinsky.

To the ILGWU's far-flung union label promotion efforts, Mrs. Matheson will bring the verve and dynamism that has characterized her union activities for close to three decades. In Wilkes-Barre, Mrs. Matheson played a prominent role in the area's labor, community, and civic activities.[60]

Nearly concurrent with Min's relocation to New York, forces beyond the control of the ILGWU would begin to permanently alter the status of the apparel industry and its workers throughout the anthracite region, Pennsylvania, and the United States.

7

Importing Apparel and Exporting Jobs

They got more sophisticated in Guatemala.

—Lois Hartel, 1996

Throughout your life you will meet people greater and lesser than yourself. It
will be a hollow victory if you rise to the top at the expense of others. Treat
them with the dignity they deserve, as you yourself would wish to be treated.

—John J. Pomerantz, chairman, Leslie Fay, Inc., 1992

We don't need companies like Leslie Fay in this valley, in this
country, or in this world . . . who believe it is okay to seek out
slave labor because slave labor isn't legal here.

—John Watson, 1994

By 1962 the three locals of the ILGWU's Wyoming Valley District consisted
of 9,400 members employed in 168 unionized factories, a number that
would grow to an all-time high of over 11,000 within a few years. The district was the second largest ILGWU stronghold in Pennsylvania, next to the
Allentown/ Reading District. It also included the second largest share of
the 89,000 members of the union's multistate Northeast Department, com-
prising 48 locals and 13 district councils stretching from northern New
England to Wilmington, Delaware (the Northeast Department did not

include urban Philadelphia, which was divided into three districts/joint boards; see Tables 3 and 4).

Of the Wyoming Valley members, 5,100 were employed in dress manufacturing, 1,100 made corsets and brassieres, 1,000 produced blouses, 760 made sportswear, and the remainder were scattered among the factories that made children's clothing, women's suits, and garment accessories.

Despite the district's remarkable growth, events in early 1963 caught many union members by surprise. Min Matheson accepted the position of director of the Union Label Department at ILGWU headquarters, at 1710 Broadway, near Manhattan's Times Square. Her new role was to create and promote a new label design, and she was so successful that by the 1970s the ILGWU label had become familiar to millions of American consumers who heard the song and saw the "Look for the Union Label" print and television ads. Bill retained the title of education director for the district and still found time for other union-related work. Wyoming Valley members protested Min's move, but the District Council ultimately gave its approval. According to one newspaper account:

Table 3 Membership in Pennsylvania ILGWU District/Joint Boards, 1962

Board	Membership
Reading/Allentown District	11,121
Wyoming Valley District	9,362
Philadelphia Dress Joint Board	8,877
Philadelphia Knitgoods District	7,591
Easton District	7,055
Philadelphia/South Jersey Joint Board	6,537
Scranton/Sayre District	5,811
Shamokin/Sunbury District	5,482
Central/Western Pennsylvania District	5,218
Johnstown District	4,316
Pottsville District	3,080

SOURCE: *Report of the General Executive Board to the Thirty-first Convention of the ILGWU*, May 23, 1962 (New York: ABCO Press, 1962), and Northeast Department district membership statistics provided by Lloyd Goldenberg, UNITE! Auditing/Accounting Department. In 1962 the Northeast Department of the ILGWU consisted of 51 locals with 88,740 members in Pennsylvania (excluding Philadelphia), Massachusetts, Rhode Island, Delaware, upstate New York, Vermont, and northern New England.

Table 4 ILGWU Membership, International and Wyoming Valley District, 1946–1962

Year	International	Wyoming Valley District
1946	379,197	2,258
1948	422,150	2,915
1950	431,200	4,267
1952	430,830	5,497
1954	440,650	7,558
1956	450,802	8,312
1958	442,901	7,757
1960	446,554	9,139
1962	441,138	9,624

SOURCE: UNITE! Research and Auditing/Accounting Departments, Census Records, 1950–62, courtesy of Lloyd Goldenberg; and *Report of the General Executive Board to the Thirty-first Convention of the* ILGWU (New York: ABCO Press, 1962).

In view of the storm in the wake of the announcement that Mrs. Min Lurye Matheson had been named Director of the Union Label Department of the union with Headquarters in New York, the farewell dinner for her and her husband, William [*sic*] Matheson, education director for the union, who will accompany her, should be a memorable affair. Never has greater Wilkes-Barre witnessed a demonstration such as the transfer of Mrs. Matheson evoked. The reaction has not only attested to the loyalty she commanded from the rank-and-file of the union, but the esteem in which she was held in the community after two decades of service. It was only after Matheson had appealed to associates to put their loyalty to the union above all else that her transfer was approved.[1]

In advance of her departure, more than seven hundred people attended a testimonial dinner for the Mathesons. Rank-and-file members and representatives from the union's hierarchy joined with elected officials, judges, clergy, heads of local colleges and businesses, and friends who knew and worked with the guests of honor. A newspaper editorial commented:

Mrs. Matheson is quite a woman. She has been controversial in union activities but, at the same time, she held the affection of thou-

Min and Bill Matheson at the ILGWU holiday party, 1960. (Courtesy of Stephen N. Lukasik, Lukasik Studio.)

sands of garment workers in the area whose standard of living was raised and whose working conditions were bettered through the concentrated and devoted efforts of Mrs. Matheson on their behalf.

Never did we see a woman work so hard at a picket line in the initial disputes as Mrs. Matheson endeavored to build the Garment Workers Union in the region. She became so enthusiastic for the cause of the workers. She became overly vigorous at times. To say she didn't make any enemies in the battle would be naive—because "enemies" were never a problem for the local head of the garment workers. Mrs. Matheson would be on the picket line ordering, demanding, blocking for an entire day. And, that night she would go out of her way to raise funds for some needy family or spark some worthy cause to success.

When it came to the good of the area, Min was smart, vigorous and determined. We could list a hundred occasions when we witnessed Mrs. Matheson do good for the region. But space will not permit.[2]

She was succeeded by Paul Strongin, an ILGWU manager from Local 93 in Reading, Pennsylvania. He was followed by Sam Bianco, manager of Pittston Local 295, and then by Lois Hartel, also a Pittston manager.

In 1972 both Min and Bill retired from their posts at ILGWU headquarters. They returned to the Wyoming Valley shortly thereafter to live close to family and friends. Min admitted that her personal and emotional attachments to the place where she had built a union and raised a family beckoned her: "I could have retired anywhere, New York, Chicago . . . but I always felt that this is home."[3] For the Mathesons retirement did not necessarily mean relaxation and leisure. They were not ones to withdraw from activist lives.

Their return to the area coincided with the most destructive natural disaster ever to strike Pennsylvania. Four days of torrential rain from Tropical Storm Agnes in June 1972 caused flooding and property destruction totaling $1.5 billion. Pennsylvania was declared a disaster area. Damages in the Wyoming Valley, the hardest-hit area in the state—thousands were rendered homeless—were estimated at $1 billion. Sixteen municipalities suffered damage. Fifty thousand people were temporarily unemployed and hundreds of businesses, schools, churches, and other community institutions were destroyed. Residents faced a monumental recovery that would take years. Min, with the assistance of Bill and their daughter Betty, founded the Flood Victims' Action Council (FVAC) to provide residents with a voice in dealing with local, state, and federal recovery agencies. When the federal and state governments allocated over $1.1 billion for rebuilding, the FVAC was highly vocal in ensuring that ordinary residents received their fair share of assistance. "I'm an atheist. But sometimes I think someone up there is doing something," was her assessment of her return to Pennsylvania just before the flood.[4]

During retirement Min provided advice, assistance, and organizing skills to the ILGWU and the broader labor movement. She received several awards, including the Pennsylvania Labor History Society's Mother Jones Award and the Humanitarian Award conferred by the American Federation of State, County, and Municipal Employees Council 87. As time passed, declining health kept her confined to Betty's home in Kingston. She died in Wilkes-Barre's General Hospital on December 11, 1992. Debilitated by Alzheimer's disease, Bill had preceded her in death in 1987. Min's obituary noted:

> At a 1977 testimonial dinner 255 people honored Matheson includ-
> ing Congressman Daniel J. Flood who, at that time, related how
> Matheson had dedicated "her entire life to helping people." He
> said that she was worthy of the tribute paid to her and that it was
> "long overdue." At the testimonial, Congressman Flood referred to
> the group [Flood Victims Action Council] as "a fearless group of
> comrades" and to Mrs. Matheson as one of "the dearest and oldest
> supporters" he had.[5]

Shortly after her death, the ILGWU's *Justice* profiled Min and her long career:

> Min Matheson's war against the Pennsylvania underworld was not
> the usual organizing drive. Nonunion employers ran away to Penn-
> sylvania, paying the wives and daughters of unemployed coal min-
> ers as little as $16 per week. New York's gangsters followed pro-
> tecting the turf of nonunion shops. Into this battlefield came Min
> Matheson. The intimidation tactics of gangs like Murder, Inc. were
> often fatal. But Matheson fought back with a weapon she figured
> they weren't prepared to cope with: women. Matheson had broken
> the spine of the gangsters' resistance. Freed from their menace, she
> signed up all but a handful of the 150 or so garment shops, includ-
> ing all of those once "protected" by the mob.[6]

Joseph Williams, a Wilkes-Barre city councilman in the 1970s, felt that "there should be a statue of Min Matheson on Public Square in Wilkes-Barre for all she has done for this area."[7] A newspaper editorial opined that "the memory of Min Matheson should be a permanent part of history."[8] And in September 1999 the Pennsylvania Historical and Museum Commission, the Wyoming Historical and Geological Society, and the Pennsylvania Labor History Society joined with UNITE! to dedicate a state historical marker to Min Matheson on Public Square. The dedication cere-

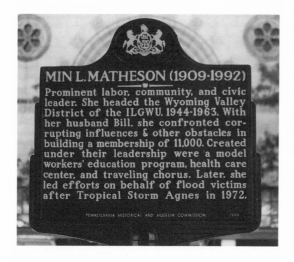

MIN L.MATHESON (1909-1992)

Prominent labor, community, and civic leader. She headed the Wyoming Valley District of the ILGWU, 1944-1963. With her husband Bill, she confronted corrupting influences & other obstacles in building a membership of 11,000. Created under their leadership were a model workers' education program, health care center, and traveling chorus. Later, she led efforts on behalf of flood victims after Tropical Storm Agnes in 1972.

PENNSYLVANIA HISTORICAL AND MUSEUM COMMISSION   1999

The historical marker commemorating the achievements of Min L. Matheson in Public Square, Wilkes-Barre. (Courtesy of George Zorgo, Zorgo Printing Service, Inc.)

mony, at which the union's chorus performed, drew a crowd of 350 people.[9]

Min's reputation remains the stuff of legend in ILGWU circles. Jay Mazur, the union's former president, said in 1998:

> I knew Min Matheson very well. I knew her until the day she died and had high regard for Min. She was tough. She had strong feelings about the union and labor movement. I spoke to her before she died and Min stuck with the union until the end.
>
> She was a leader in a union dominated by men—one of the few women [leaders]. In those days there weren't many women in leadership. She was strong. She was good for the union, ran a tough district in Pennsylvania and had to deal with all kinds of difficulties, both employers and other elements—define them in your own way—that were not favorable to the union. She was somewhat abrasive to union leadership, and kind of independent.[10]

The post-Matheson era has been fraught with struggles and problems for the garment workers and the ILGWU in the Wyoming Valley; the same struggles and problems that have affected garment workers and the ILGWU

across Pennsylvania, New York, and the United States. Some of the Mathesons' initiatives—such as community involvement, workers' education
programs, the chorus, and political activism—have continued. Elections
and candidates still draw considerable rank-and-file and retiree participation. Every spring the chorus—now consisting mainly of retirees—performs
for charitable causes in northeastern Pennsylvania. In another example of
its continuing community involvement, the union co-sponsored the hun-
dredth anniversary of the 1897 Lattimer Massacre. In Lattimer, near Hazleton in southern Luzerne County, nineteen striking Eastern European coal
miners lost their lives to the bullets of Luzerne County sheriff's deputies in
one of the most significant acts of labor violence in American history.[11]

   However, the remarkable growth experienced by the garment industry
and ILGWU in Pennsylvania's anthracite region and throughout the United
States in the post–World War II era is now but a memory. Rather than
organizing more factories, the ILGWU has become concerned with its own
survival. Membership has dwindled as production has moved overseas.
Union locals and districts have become extinct or have consolidated. Declining membership forced the union to close Unity House in the summer of
1989. The ILGWU merged with another apparel workers' union in 1995.

   In the anthracite region severe membership losses have prompted con-
solidation among districts and locals. The pressures of modern health care
economics brought the Wilkes-Barre health center to an end in 1986. And
plans called for the ILGWU headquarters on South Washington Street in
Wilkes-Barre—home to the health center since its founding—to succumb to
the City Redevelopment Authority.

   These dramatic changes have been inextricably linked to larger trends
that have affected the U.S. garment industry since the late 1960s. Indeed, what
has happened to garment workers and the ILGWU in Pennsylvania represents
what social scientists and historians have come to call "deindustrialization."[12]

Garment Imports in the American Marketplace

Like other manufacturing sectors of the U.S. economy—from television
sets to automobile components to children's toys—over the last few

"Fashion Goes to Broadway," the ILGWU's float in the Spring Fashion Parade, New York, ca. 1965. (Courtesy of Kheel Center Archives, Cornell University.)

decades, the growth of overseas manufacturing of apparel to be sold in the United States has been phenomenal. American garment workers, like their counterparts in steel, autos, and electronics, have been "downsized," unemployed, underemployed—in short, deindustrialized—as the U.S. economy has become increasingly oriented toward services at the expense of manufacturing. Clothing was among the first consumer products to have its production base shifted offshore. The loss of domestic jobs and the importation of clothing made overseas has posed unprecedented problems for the American labor movement.

During the post–World War II era the ILGWU enjoyed remarkable growth as a result of economic expansion, the generally favorable public attitude toward organized labor, and its ability to organize large manufacturers, jobbers, and contractors. The union's strategy was to organize from the top down as well as from the bottom up. That is, it struck nonunion contractors and forced them to sign an agreement, then put pressure on jobbers and large manufacturers to deal exclusively with those union contractors.

To be successful, the strategy required people like Min and Bill Matheson to agitate and organize. Ever in search of low-cost labor, jobbers contracted with runaways in other areas such as the American South. But it had become increasingly difficult to hide from the ILGWU so long as the industry maintained a strong domestic presence.

As more and more contractors signed union agreements, overall ILGWU membership grew from 320,000 in 1945 to 445,000 a decade later. By the late 1960s the union reached a peak of 470,000 members. However, it soon began a downward spiral, and it has not recovered. Its ultimate decline can be traced directly to a familiar problem.

For years, a runaway or outside shop had come to mean mainly a contractor who produced apparel away from Manhattan. By the late 1950s and 1960s, however, the terms began to take on entirely new definitions. They referred to the growing number of firms that contracted apparel production beyond U.S. borders. The former ILGWU assistant vice president Gus Tyler explained:

> There was, however, one crucial difference between the latter-day transoceanic runaway and the earlier prototypes. When a firm . . . fled the union in Manhattan and had crossed the waters in Brooklyn, and later crossed the Delaware River into Pennsylvania, and still later crossed the Mason-Dixon line into Mississippi . . . the union could follow with organizers or a federal minimum wage law. But in the case of the new runaway to Japan, the union could not use the traditional methods.[13]

What accounted for this new type of runaway? Other than the obviously lower wages paid in Caribbean, Central and South American, African, Asian, and Pacific Rim countries, savings also accrued from the avoidance of benefits such as health care, vacation and sick pay, workers' and unemployment compensation insurance, and workplace safety measures.

Garment manufacturers have been welcomed by the governments of developing nations. Apparel provides part of the answer to their search for non-capital-intensive industries to spur their economies. For example, in the aftermath of World War II, Japan embraced apparel manufacturing as a relatively easy way to provide employment and stimulate the devastated postwar economy. Scarves were among the first Japanese products manufactured in large quantities, then exported to and sold in the United States at healthy profit margins. The practice provided an incentive for jobbers and contractors to begin moving production overseas.[14]

By 1975, when the pace of imports was beginning to surge, the average U.S. garment worker earned about $3 per hour while Pennsylvania garment workers earned an average weekly wage of between $90 and $110. A person doing the same job in Hong Kong received 62 cents per hour; in Korea and Singapore, 27 cents; and in Haiti, 18 cents. By the mid-1980s, when overseas products permeated the domestic marketplace, American garment workers earned an average of $5.75 per hour, while workers earned an hourly wage of 16 cents in the People's Republic of China, 57 cents in Taiwan, $1.18 in Hong Kong, and 63 cents in South Korea. Nearly 70 percent of all garments sold in the United States were imported from those countries.[15]

Pressure by U.S. retailers for ever-cheaper garments proved to be another factor behind the growth in the overseas apparel industry. Companies argued that in the highly competitive marketplace, consumers desired the best bargain, and to survive retailers must meet consumer demand. Clothing for women and children produced anywhere but the United States and Canada was inevitably less expensive. The reason for the difference in price between U.S.- and foreign-manufactured apparel, merchants acknowledged, lay mainly in the cost of labor, not necessarily in the fabric, design, or quality of the finished product. Therefore, in theory at least, consumers' best pocketbook interests lay in apparel produced overseas. To meet those interests, some retailers bypassed jobbers and large manufacturers altogether and contracted directly with foreign producers or established their own factories overseas.

Questioning the notion that the marketplace indeed motivated sellers to stock their racks with less expensive foreign apparel and pass along savings to consumers, Sol Chaikin commissioned a study after he took office as the ILGWU president in 1975, which he publicized as "A Tale of Two

Bras." The union's research office compared U.S.-produced garments with their foreign competitors on the basis of such factors as labor cost, quality, and sale price. Chaikin reported the findings before the Subcommittee on Trade of the House Ways and Means Committee. Holding up two bras purchased at Macy's in New York, Chaikin pointed out that they were identical in pattern, design, and quality of workmanship. They were identical, too, in price. Customers could purchase the bras for $6.50 each. In fact, the bras differed only in their point of origin and in the amount paid to the person who made them. One was made in the United States, the other in Mexico, where garment workers earned about a quarter of the U.S. wage. Clearly, in this instance, consumers had not benefited from the supposed low cost of the garment produced in Mexico.[16]

Assuming that perhaps the identical prices of the bras were an anomaly, Chaikin asked the Research Department to conduct the same type of study for men's and women's shirts, blue jeans, jackets, and sweaters. The findings were the same: the foreign-manufactured products—mainly from Hong Kong and Taiwan—differed not at all from those made in the United States except in the wages, monetary and social, paid to the worker. The price paid by consumers at retailers such as Hecht's and Sears was identical.

In the union's view, the retailers' arguments were flawed. Even if Americans were demanding less expensive overseas apparel, sellers were not obliging. True, they were purchasing clothing from Mexico, Hong Kong, Taiwan, and elsewhere at a lower cost than comparable U.S.-produced apparel, but the customer received no price break. Savings from lower-cost foreign garments accrued to merchants and ultimately their shareholders. In effect, retail corporations outmaneuvered the customer, jobber, manufacturer, contractor, and worker. As Gus Tyler further explained: "By the 1980s the retailers decided that they knew best what the customer wanted. The real boss was the retail outlet, usually a chain of department stores. It had no contract with the garment union. It took no responsibility for what workers were paid or for working conditions. It was the real employer with none of the obligations of an employer."[17]

Despite what the ILGWU considered to be clear evidence that, in the 1970s at least, American consumers had not benefited at the cash register from foreign-made apparel, imports grew at enormous rates. From the early 1960s to the early 1970s, the growth in imported women's dresses alone—a mainstay of production in many of Pennsylvania's garment factories and *the* product of Wyoming Valley District clothing factories— spelled clear trouble for apparel jobs (see Table 5).

Post–World War II U.S. trade policies also facilitated the trend. In many instances, U.S. products were barred from importation by the governments of Japan, Argentina, Australia, Brazil, and the nations of Central and Eastern Europe. One method that U.S. manufacturers relied on to gain access to these markets was to establish subsidiaries or move portions of their operations to those countries. Sol Chaikin noted:

> Many American corporations move production overseas because they cannot sell their U.S. made products overseas. It [is] difficult— sometimes impossible—for American made wares to penetrate these high trade walls. Sometimes our goods are excluded by tariffs; more often the obstacles are thousands of bits and pieces of "red tape" that make entry impossible. The [result] is U.S. investment overseas. The forms of multinational operations are protean: wholly-owned subsidiaries, joint ventures, franchises, patents, and the use of overseas producers as captive contractors for U.S. concerns.[18]

The initial trickle of foreign wares led to some federal protections. Pressure from the U.S. government resulted in Japan's 1956 agreement to cur-

Table 5 Women's Dresses Imported, 1961–1972

Year	Dresses imported (thousands)
1961	3,323
1969	21,986
1970	25,227
1971	30,324
1972	26,917

SOURCE: *Report of the General Executive Board to the Thirty-fifth Convention of the* ILGWU, May 31, 1974 (New York, 1974), p. 53.

tail exports to the United States until the end of the decade. In 1961, the Kennedy administration negotiated the Cotton Textile Agreement, regulating the rate of growth in U.S. imports of apparel from developing nations where cotton was the primary raw material used in production. This agreement, renewed in 1967, served as the basis for bilateral agreements negotiated between the United States and exporting nations. The agreements capped the rate of growth of clothing imported to the United States at 5 to 7.5 percent per year.

By the early 1970s, the Multi-Fiber Agreement (MFA) replaced the Cotton Textile Agreement to cover imports from developing nations (the term "multi-fiber" refers to both cotton and synthetic material). By the mid-1970s, the United States was signatory to MFAs with more than thirty countries, limiting annual rates of import growth to 6 percent. Throughout the late 1970s and 1980s, renegotiated MFAs sought to curb imports further by imposing new and usually lower rates of growth. Though it was initially thought to be a short-term measure, the arrangement has been continually renewed. Currently about thirty-three countries have signed the MFA. Although the ILGWU "welcomed [these measures] as a forward step," more often than not they proved to be full of loopholes.[19]

U.S. tariff legislation provided scant protection to the domestic industry and ran contrary to the goals of MFAs. Item 807, for example, established lower customs duties on finished products returned to the United States, a factor essentially ignored in negotiated trade agreements. An increasingly common practice has been for domestic apparel producers to supply foreign contractors with parts of garments cut in U.S. shops. Once assembled in overseas factories, the garments are then returned for distribution in U.S. retail markets, benefiting from less expensive labor and low import tariffs.

Import quotas proved inadequate as well. They were too flexible, as Gus Tyler noted: "If a country did not fill its quota for skirts one year, it was allowed to transfer that quota of skirts to the next year. Played skillfully, such a loophole allowed a country to build up enough exports of a given product at a given time to flood the American market and wash away many American producers."[20]

Another problem was "transshipping." Nations whose exports were regulated by quotas shipped finished garments to countries where quotas did not apply or to countries that were regulated but whose total garment exportation fell short of their yearly allotment. From there the garments were shipped to the United States. Trade agreements also allowed rates of growth for imports to exceed the rate of growth in the U.S. marketplace. In years when the U.S. apparel market grew by 1 percent, for example, negotiated agreements permitted imports to increase by 6 percent or more.

As a result of the drive for inexpensive labor, the open arms of foreign governments, pressure from retailers to drive down costs, and free trade policies, the U.S. garment industry has experienced a substantial restructuring that began in the 1950s and continues to the present. Not only has the ILGWU been hurt by the resulting job losses, but equally troublesome are the sweatshop conditions in factories from Guatemala to Mexico to Sri Lanka to Vietnam.

A 1998 study of garment manufacturing in Mexico by the Highlander Research and Education Center of Tennessee revealed that in the town of Tehuacán—about two hours south of Mexico City—over 5,000 workers produced brand-name denim products in repressive and exploitive conditions for fractional wages. Twelve hours or more per day were common, as was child labor. Attempts to organize workers were strongly resisted by factory owners. In the Commonwealth of the Northern Mariana Islands— a U.S. territory—factory owners lured workers from Third World nations with the promise of U.S. green cards. Called "guest workers," people from Bangladesh, China, Nepal, Sri Lanka, and the Philippines have outnumbered the 27,000 U.S. citizens in recent years. The AFL-CIO found that they have worked in sweatshops that regularly overlook the island's mandated $3.05 per hour minimum wage. They likewise have lived in factory-provided barracks behind razor wire fences.[21]

Not only have the sweatshops that the ILGWU struggled to organize in Manhattan and Pittston emerged in overseas nations, but they have also reappeared at home in New York, Los Angeles, and other cities. Like their counterparts at the turn of the twentieth century, some contemporary contractors have taken advantage of immigrants (many of them illegal) to produce apparel and have disregarded U.S. labor laws. Indeed, the working conditions that the ILGWU struggled to eliminate earlier in the twentieth century have returned.[22]

Fighting for the Union Label

The Domestic Impact of Imports

In 1955 apparel imports amounted to a mere 3 percent of the total U.S. market. By the mid-1960s they accounted for 12 percent. In the mid-1970s the figure stood at 31 percent; by the early 1990s, sales of imports claimed over 60 percent of the total market, and the number was still growing as the twenty-first century began. Put another way, in the early 1960s, about 5 out of every 100 garments sold in the United States were made in foreign nations. By the 1990s the number climbed to over 60 of every 100. And by 2000 it was estimated that 88 of every 100 garments on the racks of U.S. clothing retailers were made overseas.[23]

The total number of women's and children's apparel workers nationwide stood at over 660,000 when David Dubinsky retired as ILGWU president in 1966. Of these a total of 455,000 were organized by the ILGWU. Twenty years later, when Jay Mazur took office as the union's president, employment fell below the half-million mark and ILGWU membership had declined to 196,000. By the mid-1990s, ILGWU membership plummeted to 125,000 as domestic employment continued its downward slide. On average, the United States has lost 40,000 combined apparel and textile jobs annually since 1979, or about 3,300 jobs each month.[24]

Pennsylvania is no longer an important center of clothing manufacturing. Employment peaked in the late 1960s, when about 180,000 Pennsylvanians made their living in some aspect of the apparel-making industry. Throughout the 1970s and 1980s the industry felt the effects of long-term permanent decline. At the dawn of the twenty-first century the Pennsylvania Department of Labor and Industry estimated that about 30,000 workers were employed in some aspect of the industry statewide. The agency further estimated that the industry will continue to loose more than 5 percent of its workforce annually (see Table 6).[25]

In 1968 the Philadelphia-area women's and children's apparel industry alone boasted 25,000 ILGWU members in 550 shops. By 1977, the number fell to less than 17,000 in fewer than 300 factories. From 1988 to 1992 Philadelphia lost another one-quarter of its apparel jobs along with an additional 16 percent of its manufacturing firms.[26]

Table 6 Number of Persons Employed in the Apparel Industry in Pennsylvania and in the Scranton/Wilkes-Barre Metropolitan Statistical Area, 1956–1999

Year	Pennsylvania[a]	Scranton/Wilkes-Barre MSA[b]
1956	165,000	27,650
1960	166,100	26,000
1964	169,700	25,400
1968	174,500	27,700
1972	162,600	26,300
1976	137,200	24,100
1980	123,189	20,275
1984	120,275	18,241
1988	82,500	12,315
1992	63,510	8,438
1996	42,996	3,595
1999	33,000	1,300

[a]*1956 to 1968:* Pennsylvania Department of Labor and Industry, Bureau of Employment Security, *Pennsylvania Employment and Earnings* 1, no. 1 (January 1956); 6, no. 1 (January 1960); 10, no. 1 (January 1964); 14, no. 1 (January 1968). *1972–1999:* "Units and Employees in the Apparel Industry," a report prepared for the authors by the Pennsylvania Department of Labor and Industry, Center for Workforce Information and Analysis.

[b]*1956–1972:* Pennsylvania Department of Labor and Industry, Bureau of Employment Security, *Scranton Labor Market Letter* 11, no. 1 (January 1956); 15, no. 1 (January 1960); 19, no. 1 (January 1964); 23, no. 1 (January 1968); 27, no. 1 (January 1972); and *Wilkes-Barre/Hazleton Labor Market Letter* 11, no. 1 (January 1956); 15, no. 2 (February 1960); 19, no. 1 (January 1964); 23, no. 1 (January 1968); and 27, no. 1 (January 1972).*1972–1999:* "Units and Employees in the Apparel Industry," a report prepared for the authors by the Pennsylvania Department of Labor and Industry, Center for Workforce Information and Analysis.

According to Lois Hartel, district manager of the ILGWU's Northeastern Pennsylvania District Office from the early 1980s to 2000:

> Imports. Imports. Imports. Imports were a threat in the 1960s. Then they got more sophisticated in Guatemala. The last twelve to fifteen years it really started increasing. When I came on staff in the 1970s, people were already saying how bad it was. Overseas competition got greater and greater. Years ago the dress industry is what our local area was basically doing. We always felt protected. The dress people always kind of felt more secure, that the jobs would stay in

this country. But as time went by that became less and less true. Because during that time they were learning overseas how to do a rapid response. How to get things done and ship them back in a hurry. During all that time we were very fortunate in the Wilkes-Barre area because we had Leslie Fay. We always felt very comfortable and safe that they'd always be there. But through the course of events—not just imports, but they got caught in a little cooking-the-books scandal—it pushed the rest of the domestic jobs overseas. They were trying to make things look good for the shareholders. They went into bankruptcy.[27]

In accord with the views of the ILGWU, Hartel saw the reasons behind the flight of jobs as economic: "The incentives to go overseas certainly include cheap labor. Certainly it is a lot easier to do business as far as taxes. They do not pay health care. They do not have workers' compensation insurance. They certainly don't have unemployment insurance for those sweatshops overseas. In the long run they [manufacturers] are lining their own pockets. Especially companies that are on the stock market. They jump at it!"[28]

Active membership in what was the Wyoming Valley District of the ILGWU has declined from a peak of 11,000 in 1968 to fewer than 400 today. In response to the decline, in 1986 ILGWU locals in Hazleton merged with their Wyoming Valley counterparts to form the Hazleton/Wyoming Valley District Council. By the mid-1990s Scranton-area locals were consolidated as well, forming the Northeastern Pennsylvania District Office. According to Hartel, "My last census was in 1996. We had about 3,000 members (active and retired) in the entire Northeastern Pennsylvania area. The numbers in the Wyoming Valley area declined substantially in the past few decades. It is hard to believe that when I went to Pittston in the mid-1970s I had 3,000 members in one local!"[29] (See Table 7.)

As a new century dawned declining membership dictated additional structural changes to the garment workers' union. In 1995 the ILGWU merged with a major textile and apparel union to form a new entity—a subject more fully discussed in the Epilogue—and three joint boards were created to represent about 20,000 Pennsylvania workers: the Philadelphia Joint Board with 4,000 members; the Mid-Atlantic Joint Board with

Importing Apparel and Exporting Jobs

Table 7 ILGWU Membership, 1964–1998

Year	Total[a]	Wyoming Valley District[b]
1964	442,318	10,279
1966	455,164	10,901
1968	455,022	11,179
1970	442,333	11,006
1972	427,368	9,634
1974	404,737	8,900
1976	365,346	7,990
1978	348,380	7,374
1980	322,505	7,187
1982	282,559	6,508
1984	247,570	6,055
1986	196,000	4,817
1988	153,000	3,688
1990	145,000	3,570
1992	133,000	2,950
1994	125,000	1,839
1996	245,000[c]	1,658
1998	217,000[c]	396

SOURCE: Union census records provided by UNITE! Auditing/Finance Department, New York.
[a] United States and Canada.
[b] Active members in Locals 295 (Pittston), 249 (Wilkes-Barre), and 327 (Nanticoke). From 1946 to 1986 these locals constituted the original Wyoming Valley District.
[c] Data reflect the 1995 merger of ILGWU and ACTWU.

8,000 members; and the Pennsylvania, Ohio and South Jersey Joint Board, with 8,000 members. Most were employed in warehousing, health care, and equipment manufacturing while only a small percentage were employed in the actual manufacture of apparel.

Garment factory owners, like garment workers, have also seen dramatic shifts in the industry. In the late 1960s the Pennsylvania Department of Labor and Industry classified over 2,000 facilities as in some way related to the production and shipment of apparel. Three hundred fifty were located in the Wilkes-Barre/Scranton metropolitan area. By 2000 the department reported that three-fourths of those facilities throughout the state no longer existed; a few dozen remained in the northeastern quadrant.[30] As the Pittston manufacturer Leo Gutstein explains:

I've been president of a trade association in northeastern Pennsylvania for ten years. When I took over we had two hundred members. I currently have seven. You can see the erosion that has happened in the industry. Pittston alone had, on Main Street, forty factories at one time. Today [1997] we are sitting here in the only one that is left. My father started this business with ten or twelve operators, then it had about twelve hundred people working in it. Today I have less than one hundred.

My competition is nonunion. My competition is foreign labor. My competition is facilities that are uncontrollable by the union. [Production of dresses] has even moved back to the Lower East Side of Manhattan with immigrant labor. Apparel is made in China, Korea, Latin America. The union can't control the fact that these people work sixty to seventy hours a week for below minimum rates.

Being in an international industry, I've had the fortunate or unfortunate experience of going to the Orient and seeing what my competition does there. For them to work sixty or seventy hours a week is not abnormal. Workers were afraid to raise objections. After spending time in shops in Hong Kong, Korea, and Taiwan, I realized that they really didn't have a skill level that was anything better than my employees had. In fact, I believe that my employees were much more productive and have a better skill level. The difference was that they received one-tenth the pay that my workers had!

I've also had the experience of spending some time in the Caribbean and the skill level is terrible. But the pay again is something that is very difficult to compete with. In fact, I was in the Dominican Republic and [set up] a factory for a while. But it turned out to be a bad business judgment on my part. So my partner and I took our losses and left.[31]

While Pennsylvania's apparel industry was already in steep decline as the last decade of the twentieth century dawned, ensuing events in the district once managed by Min Matheson would epitomize the desperate struggle of apparel workers to keep their jobs from going overseas. The ILGWU's battle with Leslie Fay, Inc., proved to be one of the most dramatic national examples of the strategies employed by apparel manufacturers and the reactions of workers and organized labor in the postindustrial economy.

"Keep Leslie Fay in the U.S.A."

In 1946 Fred P. Pomerantz relocated from New York to the Wyoming Valley and opened a dress factory, which he named after his daughter. Fred and his twenty employees set out to produce petite women's fashions in a two-story building in Plymouth Borough. Over the next twenty years Leslie Fay, Inc., opened a major manufacturing and distribution plant on Highway 315 in Plains Township, near Wilkes-Barre, hired 1,500 workers, and saw its annual sales approach $100 million. By the late 1960s Leslie Fay was ranked among the nation's premier manufacturers of women's clothing and was recognized as a pioneer in petite fashions. The company was highly regarded as a local employer as well. Indeed, in 1958 Fred Pomerantz was presented with a maroon Rolls-Royce, paid for in part by ILGWU members, who had had a contract with the company since the late 1940s.

In 1991 the company's sales hit nearly $900 million and its new chairman, John J. Pomerantz, Fred's son, dedicated its new administrative headquarters at the Wyoming Valley's Hanover Industrial Park. The company employed 4,500 internationally and 1,800 locally, with a total local payroll of $40 million annually. In May Leslie Fay and the ILGWU agreed to a new three-year contract granting pay and benefit increases that brought the average hourly wage of production workers to about $10.

The esteem in which the company was held transcended two generations. When the younger Pomerantz was invited to deliver the commencement address at Wilkes University in 1992, he boasted that his board had recently authorized "the development of a multi-million-dollar, state of the art, dress manufacturing facility here in Wilkes-Barre . . . ensuring the continued employment of hundreds of Leslie Fay personnel in the community."[32] Pomerantz received warm applause from the large crowd. It appeared as though the company were planning to continue its stake in Pennsylvania's economy.

Early in 1993, however, auditors discovered accounting irregularities at Leslie Fay totaling $81 million. Among other problems, the company had apparently understated expenses. Despite the fact that the company reported sales of $675 million in 1993, it had difficulty paying its bills and was forced to file for bankruptcy protection in April. By late summer 1993, management announced plans to close its Julie II factory in Tuscarora, Schuylkill County, despite an agreement with the ILGWU stipulating that it would maintain its Pennsylvania facilities. The ILGWU filed for a preliminary injunction to prevent the loss of more than seventy jobs at Julie II Apparel. Leslie Fay's lawyers argued that closing the plant was simply a business decision. Citing violation of its contract with the ILGWU, an arbitrator ruled in the union's favor and the jobs were saved. It soon became clear, however, that the ILGWU's problems with Leslie Fay had just begun.[33]

By the spring of 1994, the company announced plans to reduce domestic production and contract work to nonunion factories in Guatemala. Many of the workers at Wilkes-Barre's "main plant," as it was known, had their weekly hours reduced from 40 to half or fewer. Some were furloughed. Workers at other Leslie Fay factories and contract shops in the area were affected as well. Those at Ricky Fashions, Kingston Fashions, Throop Fashions, Andy Fashions, and Pittston Fashions faced cutbacks. Facing bankruptcy, Leslie Fay had begun to reneg on the ILGWU contract.

Rumors abounded that the company planned to move all production to Guatemala when its union agreement expired on May 31, 1994. The company confirmed the rumors in late March, when it sat at the negotiating table with ILGWU officials. According to Leslie Fay negotiators, losses of nearly $100 million in 1993 left them no choice but to close all U.S. facilities. The production would indeed shift to Guatemala. Leslie Fay's announcement prompted the ILGWU to mobilize its most determined campaign since the 1958 general dress strike. The situation also galvanized people throughout Pennsylvania in a broad movement to save the jobs.[34]

In March and April 1994, workers organized "militant action teams" and staged peaceful demonstrations at Leslie Fay facilities. Jay Mazur and Sol Hoffman, the union's president and vice president, met with district members to mobilize support and plan for upcoming contract negotiations. ILGWU attorneys filed an unfair labor practice complaint with the National Labor Relations Board. And the union helped organize the Northeastern Pennsylvania Stakeholders Alliance, which included clergy, social agency representatives, business and labor leaders, and politicians ranging from members of the Pennsylvania General Assembly to mayors and county officials. The official position of the Alliance was that the company had an obligation to maintain jobs in the community and that their removal would seriously injure the economy and quality of life in the region.[35]

The Alliance's efforts were boosted by the support of religious and labor leaders. Bishop James C. Timlin of the Roman Catholic Diocese of Scranton and Bishop Harold Weiss of the Northeastern Pennsylvania Synod of the Evangelical Lutheran Church in America publicly opposed Leslie Fay's plans to downsize, as did William George, president of the Pennsylvania AFL-CIO. Each sent representatives to Alliance meetings.[36]

The union argued that it was "waging one of the most militant, unified counterattacks in labor movement history" and that "we're not going to the unemployment line without the fight of our lives."[37] Members vowed to remain on the job for the remainder of the contract. Though Leslie Fay would clearly profit by moving production to Guatemala and paying workers as little as 35 cents an hour, ILGWU members hoped that the force of law as well as an appeal based on public opinion and the community's economic viability would prevail. As contract negotiations continued, ILGWU members stepped up their militancy. More than fifty members traveled to the home of Leslie Fay's president, Michael Babcock, in Greenwich, Connecticut, in mid-May. They carried signs protesting Leslie Fay's plans, chanted, "Hey, hey, Leslie Fay, keep our jobs in the USA," and sang "Solidarity Forever" at the entrance to Babcock's exclusive neighborhood.[38]

Pennsylvania's governor, Robert P. Casey (a native of Scranton), and lieutenant governor, Mark Singel, also supported the union's effort. Casey wrote Michael Babcock asking that his company reconsider its plans. Moreover, the governor offered low-interest loans for equipment upgrades and $500,000 in customized worker training programs to save the jobs. Lieutenant Governor Singel took a more critical approach, noting, "I am truly angered that Leslie Fay has determined to close the doors on its production and other facilities here in Pennsylvania. I will do everything I can to try to preserve the jobs. The company's course of action is utterly indefensible."[39]

As the end of May neared, the ILGWU rejected Leslie Fay's offer to retain the 1,200 local jobs at current pay rates for one year and to provide monetary incentives for employees who chose to leave voluntarily. As of May 1995, the company would guarantee about only 200 local jobs, and these were located at its warehouse. All 2,000 of the company's domestic production jobs would be transferred to Guatemalan contractors. Rank-and-file members authorized a thirty-two-member committee to call a strike effective Wednesday, June 1. Early that morning, about twenty-five ILGWU members arrived for picket duty at the Route 315 plant, which soon became the center of activity in a strike affecting 2,000 Leslie Fay workers in Pennsylvania, New York, New Jersey, and Georgia.[40]

Before the strike, the ILGWU had commissioned a study to determine the economic impact of plant closings in Pennsylvania. According to the Midwest Center for Labor Research, while 1,200 local jobs would be lost, a ripple effect would cause another 1,800 people to face underemployment or unemployment. Lost local, state, and federal tax revenue would total $14.7 million over two years and unemployment benefits would cost $10.1 million. In the days immediately following the strike, the head of the Economic Development Council of Northeastern Pennsylvania echoed the study's findings. Arguing that a shutdown would be devastating, Howard Grossman pointed out that "it's not good for the country to allow these jobs to go overseas" and that the strike was "probably the single most critical event of substantial size to occur here since the coal mining days." Grossman's comments came on the same day that U.S. Senator Harris Wofford visited the ILGWU's picket line, telling a reporter, "I'm here simply to express my support for Leslie Fay workers."[41]

The strike continued to attract the attention of public officials and the media. On June 7 Pennsylvania Congressmen Ron Klink, Tim Holden, and Paul McHale joined Paul Kanjorski of the Eleventh District and other members of the House Subcommittee on Labor-Management Relations at public hearings in Wilkes-Barre to examine the situation. Those offering

Sen. Harris Wofford visits striking Leslie Fay workers, June 1994. (Courtesy of ILGWU.)

testimony included Leslie Fay workers, union representatives, local officials, and Pennsylvania's commerce secretary. Absent from the hearing was
John Pomerantz, who, though invited to attend, responded that it was
inappropriate for a congressional committee to hold hearings to attempt to
influence what was essentially a private matter between Leslie Fay and its
workers.

The most dramatic testimony came from Guatemalan garment workers,
who captivated listeners with tales of child labor, unsanitary conditions,
low pay, and long hours. Dorka López told the panel that she earned 43
cents for each skirt. A Leslie Fay employee, Pearl Novak, informed the
panel that the skirts to which Ms. López referred were sold by New York
retailers for $48. Yet for $50 the same retailers sold the same type of skirt
made by Wyoming Valley workers, who earned about $8 an hour. Con-
gressman Kanjorski concluded that "there is indeed exploitation in the
global market" and that "U.S. industry, by persuasion or by statute, must
be made more responsive to environmental, wage, and labor issues."[42]

Lois Hartel recalled the hearings and the testimony of the Guatemalan workers as one of the defining moments of the ILGWU's struggle with the company:

> We had a congressional hearing that Paul Kanjorski set up in Wilkes-Barre and we had Guatemalan workers testify. They told what the conditions were like. They had the attention of everybody in that hall. I think some people were on the verge of tears.
>
> Some of the children—fourteen, fifteen years old or younger—should be in school but instead they have to work. People working from dawn until dusk for terrible wages. It made you feel really bad for these people, to see them exploited like this and then to think that Americans buy the products that they make and don't think twice about it.
>
> It is all the same type of stories about sweatshops that you used to hear at the turn of the century in New York. Here we are one hundred years later locked in a battle with a company who could care less about sending jobs to places that are nothing more than sweatshops.[43]

After the hearing, Congressman Kanjorski announced that he planned to conduct a follow-up session and once again invited John Pomerantz to testify. Leslie Fay officials responded that "we will not meet in an open forum which is clearly intended to be a political event rather than a legitimate attempt at fact finding."[44]

The AFL-CIO and ILGWU called for a national boycott of Leslie Fay products and retailers who carried them. In mid-June the union called a rally at the company's offices at 1400 Broadway in Manhattan. Strikers later distributed boycott leaflets on Wall Street and at Saks Fifth Avenue and other fashionable department stores. A rank-and-file delegation also met with John Cardinal O'Connor, archbishop of the Roman Catholic diocese of New York (former bishop of the diocese of Scranton), to discuss the situation. After the meeting Cardinal O'Connor publicly expressed his support for the workers.[45]

One week later, the company threatened to withdraw from the negotiating table unless the union accepted its late-May offer. To substantiate the

Importing Apparel and Exporting Jobs

Strikers urge a boycott of Leslie Fay at the New York Stock Exchange, Wall Street, June 1994. (Courtesy of ILGWU.)

threat, it completely closed four factories in the Wyoming Valley, including the main plant. Both the company and the ILGWU reported no progress in contract talks. A federal judge ruled that the bankruptcy proceedings made the July 1993 agreement between the parties guaranteeing the mainte- nance of domestic union jobs unenforceable.[46]

The union enjoyed favorable community and political support during the walkout. On June 20, the Pennsylvania House of Representatives passed Resolution No. 351, urging the parties to resolve their differences and imploring Leslie Fay to stay in Pennsylvania. Ten thousand Catholic parishioners in Luzerne, Lackawanna, and Wyoming counties signed a "Community Letter of Concern" urging Leslie Fay to reconsider its posi- tion. National media focused attention on the dispute. CNN and other net- works and major newspapers highlighted the story. Jay Mazur told the *Wall Street Journal*, "For the ILGWU, our goal for nearly 100 years remains our goal today: Jobs with respect and dignity for garment workers."[47]

While a New York bankruptcy judge considered granting Leslie Fay executives pay raises and bonuses totaling more than $6 million—a move that drew angry protests from ILGWU members at a Manhattan courthouse

Strikers rally support against Leslie Fay in New York's garment district, June 1994. (Courtesy of ILGWU.)

and at a Wall Street rally—both sides announced that William Usery, former U.S. secretary of labor, would mediate the dispute. As members continued to walk the picket line in Wilkes-Barre, the New York judge approved an upgraded compensation package for Pomerantz, his wife, Babcock, and other company executives.[48]

On July 12, six weeks after the ILGWU had walked out, both sides announced a tentative three-year contract. Leslie Fay agreed to maintain 600 production jobs at the main Wyoming Valley plant for one year, through July 31, 1995. Both the company and the union resolved to work toward preserving the jobs for the second and third years of the contract. The company further provided a total of $2.3 million in severance pay to terminated workers, wage increases of 10 percent for those who remained on the job, increased contributions to the workers' health and welfare fund, and fully funded health insurance co-payments. Leslie Fay also consented to adopt a code of conduct prescribing fair labor standards for its Guatemalan operations. The company and union agreed to continue to work with Usery to resolve any lingering concerns, including the ultimate fate of the

600 remaining jobs. Despite the fact that, with a few exceptions, the compact did not substantially differ from the offer Leslie Fay had originally proposed in May, ILGWU members voted to approve the contract on July 13.[49]

Jay Mazur told the press that this was "a trailblazing agreement. It points the way to joint labor-management efforts to preserve American manufacturing jobs even as American companies participate in the global economy." At least one union official cautioned, however, that many details were yet to be resolved, including what might happen to 600 workers after July 31, 1995. After a thanksgiving prayer service at the main factory on the morning of July 15, the ILGWU held a victory picnic at a grove in nearby Mountaintop.[50]

All but Leslie Fay's main plant, a nearby warehouse/distribution center, and one smaller plant soon closed permanently as jobs were moved to factories in Guatemala and eventually in El Salvador.[51] The 600 retained workers remained on the job in the hope that some fruitful resolution would be forthcoming by July 1995. But none was to be had. As the summer of 1995 drew near, the company remained in Chapter 11 and made clear that its future relied in part on reducing labor costs by moving production jobs out of the United States. Leslie Fay's argument was aided by an independent evaluation that showed it returning to profitability through further restructuring and lower labor costs. The ILGWU subsequently reached an agreement with Lesley Fay for severance pay for about 400 workers. The company promised to retain its warehouse and distribution center and a small plant to produce samples and monitor quality. The 200 or so remaining jobs would be union.[52]

As the final chapter of the Leslie Fay saga unfolded, membership in the Wyoming Valley District dwindled to levels last seen at the time Min and Bill Matheson arrived in 1944. By 1995 the apparel industry in the anthracite region had become a shadow of itself; in the Wyoming Valley six unionized factories remained. The ILGWU had lost its struggle with one of the twentieth century's largest apparel employers and, indeed, with an entire industry.[53] Capital and the marketplace had won out over the voices of workers, politicians, religious leaders, labor officials, and community members.

An abandoned factory in Easton, Pa. (Photo by Martin Desht.)

A worker carries on alone in a downsized garment factory, Shenandoah, Pa.
(Photo by Martin Desht.)

Pearl Novak, a fifty-one-year member of the ILGWU and a Leslie Fay striker who lost her job, summed it up:

> People were really hurt by this. You have to realize that many of them gave their lives to this company. Now they had nothing. Pomerantz talked about "doing unto others." He did unto others, all right! As far as the workers, some found work. Some went to school to be retrained. Some are still out of work. Many, like me, were forced into retirement. We had no choice. Sad, isn't it?
>
> I was a member of the union when Min Matheson was here. It was different then. The industry was here. Now it isn't. But I think the union did all that it could have. What else could it have done? What this really comes down to is that we need to educate young people today about why it is so important to keep jobs in the U.S.A., to buy American, to stick with the union, to fight to the end like we did at Leslie Fay. What else is there to do?[54]

Epilogue

Our progress may thus be described as having succeeded in changing our
methods of work and in sweeping away all abuse. Thus, we learn from
the past while working for the future, with good hopes for onward progress.
—M. Sandler, secretary of Philadelphia Local 58, ILGWU, 1910

The U.S. garment industry has been withering for years.
—*Philadelphia Inquirer*, December 11, 1994

Over the past two decades, the landscape of industrial relations in North America
has changed dramatically. Now is the time to advance . . . into a new organizational unity.
—Merger agreement between the ILGWU and the ACTWU, February 1995

The ILGWU's strike against Leslie Fay represents one way in which the 100-year-old union has dealt with the unprecedented challenges facing American labor in the global economy. Going with tradition, the union reinforced activism with education.

For most of the twentieth century, education involved immersing workers in a culture of unionism—a large part of the mission of Unity House, for example. It also meant ensuring that the ILGWU and its members were active in the communities in which they lived and worked, as in the Wyoming Valley District's chorus and political participation. Education also incorporated opportunities for workers to learn about industrial capitalism—an idea central to the Workers' University and other education programs. And it meant enlightening members about public and social policies and the politics associated with them.

Education permeated many otherwise routine union functions. Learning became part of organizing: it remained intrinsic to the union's community involvement, and it was a driving force in efforts to build a foundation to support members' needs. In sum, education and learning could be delineated in distinct, formal processes. However, it was also part of the informal socialization and cultural processes that went along with being an ILGWU member. One could visit Unity House, for example, to attend formal educational institutes and classes; or a member could simply vacation and relax in its pristine environment, which afforded good food, entertainment, cultural programs, and the chance to socialize with others who were, for the most part, representative of the same social class.

Notwithstanding the continued erosion of domestic apparel jobs, education—both formal and informal—has remained integral to the union's mission. In the twenty-first century, however, it has broadened its focus in response to the contemporary political economy. Education, combined with activism, now reaches beyond union members, elected officials, and civic leaders to consumers, manufacturers, governments, and international labor.[1]

Activism and Education in a New Era

The ILGWU leadership—like the leadership of American labor generally—advances the view that free trade unions are essential to ensuring that the rights of workers are protected in nations that have embraced apparel making as part of an economic development strategy. That being the case, more ought to be done to promote free trade unions domestically and abroad. The union has remained vocal in support of apparel workers' interests in the United States and around the world. Such ventures have been designed to educate workers, consumers, governments, and manufacturers about unions and workers' rights. The union encourages con-

Consumer guide to decent clothes

Were these clothes made under decent conditions?

The care tag tells you how to treat the garment, but not how the worker who made it was treated. A union label is one way to know. If you don't find one, here are some questions to ask the store manager.

UNITE! 1710 Broadway. NYC 10019

☑ Do you know how the workers who made this garment were treated?

☑ Does your store have a code of conduct for all factories that make the clothes you sell?

☑ Does it forbid child labor & protect human rights? Does it specify living wages? Is the code of conduct posted in every factory?

☑ Is there an independent monitoring agency to make sure that everybody lives up to the code?

☑ **LOOK FOR THE UNION LABEL!**

UNITE! urges consumers to look for the union label. (Courtesy of UNITE!)

sumers to consider its *Consumer Guide to Decent Clothes* when they make decisions on apparel purchases; encourages boycotts of "sweated" apparel; advocates the payment of a living wage to garment workers on domestic and foreign soil; and works with national and international organizations to call attention to issues affecting workers across the globe.

For example, the ILGWU has been a long-time supporter of the International Confederation of Free Trade Unions, which has worked since 1949 to provide material, moral, and educational support to organizing efforts in other lands, sometimes in opposition to the policies and actions of national governments.

The union also works with the International Textile, Garment, and Leather Workers' Federation, whose membership consists of unions in several nations who advocate workers' rights. The union plays an active role in the National Labor Committee, an organization that calls public attention to foreign and domestic sweatshops and targets manufacturers and retailers who engage in abusive labor practices. The union participates with the American Institute for Free Labor Development in studying and promoting free trade unions and workers' rights in developing nations. Jay Mazur has chaired the AFL-CIO's Committee on International Organizing and Solidarity to assist exploited overseas workers. And the union has actively supported the AFL-CIO's Committee on Political Education (COPE) to educate officers and rank-and file members regarding public policy and political issues.

In 1996 the union joined with industry representatives and activists in the White House–sponsored Apparel Industry Partnership. This group's work culminated in the adoption of workplace codes of conduct that banned child and forced labor, granted limited collective bargaining protections to workers, and established minimum health and safety standards for apparel labor worldwide. The Partnership founded an association that manufacturers can voluntarily join subject to their agreement to adhere to the codes of conduct when they contract with overseas producers. While this is a significant step in the right direction, the overall success of the venture continues to rely on the good faith efforts of manufacturers to abide by such codes and to have their contractors do the same. It also relies on consumer education and awareness to help determine which manufacturers do and do not abide by the codes. Though some manufacturers have joined the association, the overall success of the venture remains to be seen.[2]

In addition to its advocacy of fairer treatment for workers abroad, the ILGWU has tapped into its long history of Washington lobbying to deal with contemporary labor problems. Arguing that federal laws and policies have largely been ineffective at retaining U.S. manufacturing jobs and protecting apparel workers, the union has pushed for greater government action.

The union supported the U.S. Department of Labor's efforts to curtail sweatshops in the United States' urban centers by holding retailers accountable for working conditions and vigorously enforcing the Fair Labor Standards Act, minimum wage laws, and factory health and safety standards. The union also supported the Immigration Act of 1990, which granted amnesty to undocumented U.S. immigrants—many of whom found their way into the nonunion domestic apparel industry—with the goal of securing their rights to labor representation and collective bargaining. Another policy has been to endorse the U.S. Department of Labor's Fashion Trendsetters List, which names apparel manufacturers who have agreed to follow "no sweat" practices and abide by U.S. labor standards.[3]

This federal effort was undertaken in part to remedy rollbacks of federal bans on homeworking and lackluster enforcement of U.S. labor laws under the administrations of Ronald Reagan and George H. Bush. As we saw in Chapter 1, before factory-based apparel manufacturing, garments were produced in people's homes, often on contract with a manufacturer. In the early 1940s, the federal government banned homeworking in several sectors of the apparel industry, including women's clothing, as it was nearly impossible for regulators to enforce provisions of the Fair Labor Standards Act in those circumstances. In the 1980s, the Labor Department lifted the prohibition for most branches of the apparel industry, and the decision was upheld in federal court. The ban remained only for women's dresses, and the Bush administration responded to that ban by reducing the Labor Department's enforcement budget.

The rebirth of a domestic homeworking system has led to a spate of U.S. sweatshops. Once again the doors are open for manufacturers to contract with any willing producer, without regard to working conditions or wages. One estimate indicated that by the early 1990s, some 80,000 apparel jobs in Dallas/Forth Worth, Texas, had gone to the homeworking or underground industry. In New York, some 3,000 producers were ignoring wage and safety laws, often employing undocumented immigrant labor by contract. By the closing years of the twentieth century the problem was so widespread that U.S. Labor Secretary Alexis Herman encouraged manufacturers and retailers to aid regulators in policing compliance with the Fair Labor Standards Act:

> Many companies in the American apparel industry provide good jobs, decent wages, and quality clothes, and they deserve our support. But the firms that utilize and tolerate sweatshop labor make it hard for honest, law-abiding shops to compete in the marketplace. Both industry and labor have an interest in making sure that companies do not mistreat their employees.
>
> Today, sweatshops are an ugly stain on American fashion and it is up to all of us to remove it. A recent survey in southern California indicates that contract shops that are monitored by the manufac-

turers are three times as likely to be in compliance with labor laws as shops that are not monitored. Our goal is for every garment worker to be paid correctly, and monitoring is a powerful tool for achieving that goal.

As Secretary of Labor, I have made it my mission to ensure that all employers abide by the Fair Labor Standards Act, to pay at least the minimum wage and overtime, and to comply with the child labor and homework provisions. All workers should be treated with basic dignity.

Join the Department of Labor and many of our country's finest retailers and manufacturers in the commitment to eradicate sweat-shops![4]

In another move to draw attention to the rebirth of sweatshops, the garment workers' union co-sponsored—with K-mart Corporation and the National Retail Federation—a Smithsonian Institute exhibit titled "Between a Rock and a Hard Place: A History of American Sweatshops, 1820–Present." The exhibit explored the history of American garment manufacturing from impoverished seamstresses and sweatshops to the rise of the trade union and reform movements to present-day problems with contracting. It highlighted critical events such as the 1909 Uprising of 20,000 and the August 1995 law enforcement raid on an apartment complex in El Monte, California, where seventy-two illegal immigrants had been forced to sew in captivity. A brochure titled *The Global Production Game* educated consumers on how their apparel consumption decisions affect working conditions and wages. The exhibit was available for public viewing at the Smithsonian's Museum of American History in Washington from April 22 to December 1, 1998.[5]

The exhibit focused some attention on the issue of free trade, another area in which the ILGWU has taken a critical view. In an earlier era the union argued that the federal government had an obligation to enforce import restrictions and periodically establish new limits to discourage the flood of overseas products, the continual flight of American jobs, and the exploitation of overseas labor.[6] In recent times, however, federal policy has moved in entirely new and sometimes unforeseen directions.

Epilogue

In keeping with past endorsements of generally labor-friendly politicians, the ILGWU pinned its hopes for a sympathetic national ear on the election of Bill Clinton in 1992, despite his favorable pre-election views toward free trade. The union endorsed Clinton and worked for a Democratic election victory; the alternative would mean even more peril for American apparel workers, in the ILGWU's view. Like the labor movement generally, the ILGWU had hoped to influence Clinton and his top advisers on the problems of carte blanch free trade. But their hopes were dashed when it became apparent that the Clinton administration would not waver in its support for free trade.

The pillar of Clinton's trade policy has been the North American Free Trade Agreement, or NAFTA, which established the world's largest free trade zone. Signed by the United States, Canada, and Mexico, the treaty took effect in January 1994 and gradually eliminated tariffs on goods produced and sold throughout the three countries. Unions warned of the further erosion of U.S. jobs, the likelihood of exploitation of Mexican workers, and the potential for environmental degradation. Robert Hostetter, former education director of the Central Pennsylvania chapter of the ILGWU, summed up labor's views: "With regard to NAFTA, our views are simple. Fix it or nix it. We really don't want free trade like NAFTA. But the reality is that we are going to have to live with it. If that's the case, we must make sure that certain issues are addressed, like workers' rights and environmental conditions."[7] At the insistence of labor leaders and environmentalists, trade policy makers agreed to incorporate "side agreements" to protect the environment and improve working conditions, although enforcement would remain another matter.[8]

Despite labor's resistance, NAFTA brought about the gradual elimination of trade barriers. Partisanship proved largely inconsequential as congressional Democrats and Republicans joined to approve the treaty. Indeed, both major party candidates in the 1992, 1996, and 2000 presidential elections favored free trade, as do most Washington policy makers today. In the ILGWU's view, NAFTA's enactment not only was a direct blow to the U.S. garment industry but called into question the ILGWU's longstanding alliance with the Democratic Party. "We worked very hard for Bill Clinton's election," Lois Hartel recalled.

We knew we would have bigger problems [if he was not elected]. We knew he was in favor of NAFTA but we had hoped that after he got elected we could get him educated before he put his name on the legislation. But . . . NAFTA put the icing on the cake, so to speak. We had been in a twenty-five-year struggle with imports. Spent millions and did all kinds of educational and political programs. We were trying to not even roll back the imports. We had legislation in Washington just to keep them at the levels that they were at then. Just let us have the jobs that were left in this country. NAFTA took more jobs out of the country. It resulted in job losses in my district.[9]

With NAFTA, free trade, and the gradual elimination of apparel and textile import protections once guaranteed by the 1947 General Agreement on Tariffs and Trade (GATT), the U.S. domestic garment industry faced unprecedented challenges as the twentieth century drew to a close. In December 1994 the *Philadelphia Inquirer* reported:

The U.S. garment industry has been withering for years. But many analysts say the GATT treaty, set to take effect Jan. 1, may be fatal. For more than 30 years, textiles and apparel have been among the United States' most protected industries, shielded against imports by both tariffs and by specific country-by-country quotas on the amount of goods that can be shipped to the U.S. Despite all this protection, imports have seeped in anyway as emerging nations have developed their garment industries and America has developed a firmer policy in favor of free trade. Ten years from now, under GATT, most protections will be gone.[10]

Lois Hartel commented that, in an ironic twist of fate, "it took a Democrat to finish us off."[11]

The realization that the U.S. garment industry was in a state of perpetual decline has been echoed in places hard hit by job losses—like Pennsylvania's anthracite region—for over two decades. In this area of the Keystone State, once home to King Coal, the loss of yet another major source of employment has had a severe impact. According to a Wilkes-Barre editorial:

Greed and lack of foresight destroyed this area's original industry—coal mining—25 years ago. Today, greed and lack of foresight are destroying a major industry of the present: garment manufacturing.

Just as greed once led mine owners to go for "easy coal" just below the river [a reference to the Knox Mine Disaster of 1959], greed is now leading garment manufacturers to go for easy profits overseas. Just as lack of foresight caused mining to collapse and rendered useless a huge potential store of valuable energy, lack of foresight is allowing garment manufacturing to go to foreign countries and create a ripple effect which will hurt other businesses and eventually hurt the national economy.

It is too late to do anything about the mining industry. Sadly, but surely, it is dead. But it is not too late to do something to save the domestic garment industry.[12]

Free trade, accelerating imports, domestic job losses, exploited and powerless overseas workers, the rebirth of sweatshops in the United States, the power of business lobbyists, and unsympathetic politicians have severely weakened American garment workers and the labor movement. Apparel and textile unions have joined forces to maintain some strength in collective bargaining as well as in domestic and international issues. In 1995 the ILGWU merged with the Amalgamated Clothing and Textile Workers Union to create UNITE!, or Union of Needle Trades, Industrial and Textile Employees. The merger agreement began with a statement on the unions' traditions:

On the eve of the twenty-first century, the merger of the Amalgamated Clothing and Textile Workers Union and the International Ladies' Garment Workers' Union brings together two proud and strong unions, to form a new union, rooted in a rich history but committed to innovation. We share a common commitment to resisting the exploitation of workers: organizing the unorganized, building unity out of diversity, and pioneering labor-management partnerships to better workers' lives. We undertake to form a new union, because we know that the challenges of a global economy

The UNITE! Chorus, 1999. (Courtesy of George Zorgo, Zorgo Printing Service, Inc.)

require constant renewal of our struggle and creative, far-reaching strategies.

Then the document moved to the contemporary conundrum:

> Over the past two decades, the landscape of industrial relations in North America and the world has changed dramatically. The social contract between capital and labor forged over many decades is systematically being dismantled. Today, big business knows no loyalty to community or nation and feels free to move anywhere in the world without regard for human and social consequences. Nowhere has the impact of these changes been felt more keenly than in our own industries. North American workers are in direct competition with those less developed parts of the globe as corporations rou-

tinely shift production offshore in search of ever-lower wages and working conditions, creating a downward spiral throughout the entire industry. New trade agreements, with minimal or non-existent provisions for workers' rights, seek to broaden and consolidate the hold of these transnational corporations over the terms and conditions of labor and the quality of our lives.

The corporate assault on our hard-won union achievements of the past requires us to take bold measures to guarantee past victories and address the workplace of the future. Now is the time to advance from the traditional cooperation of our two unions into a new organizational unity.[13]

As the Leslie Fay episode unraveled, Lois Hartel witnessed the union alliance:

The merger was officially voted on in 1995, July 1, at a convention in Florida, right around the time that we realized Leslie Fay was hopeless. It is something that has been talked about for many years and they never got it together. We've always worked together because we've always had the same problems. It makes sense. The Amalgamated has a lot of membership in the textile industry, which, for the most part, we were never into. So both have strengths and weaknesses. We'll work together to survive. We've both lost members.[14]

The merger and a simultaneous $10 million campaign to organize domestic and foreign workers received wide media coverage: "Two clothing workers' unions that were among the most powerful political and social forces in labor announced yesterday that they have agreed to merge," wrote the *Harrisburg Patriot*. "Each of the unions . . . has seen its membership shrink to half or less its peak size as the needle trades and textiles moved offshore in search of cheap labor. But the combined union is expected to retain significant clout."[15]

Today UNITE! has about 200,000 members, mainly in the United States and Canada. Its challenges remain formidable, as Jay Mazur explained:

Organizing today means reaching across borders, from the tip of Canada to the tip of South America. It means reaching across the ocean to join cause with workers in other countries. The only way to raise living standards for some workers is to raise standards for all workers. It's no longer good enough to fight the boss in one town, in one state, or even one country or continent. Today, many of our companies are global; most of our industry is global. It's time to take a stand and take control of our future.[16]

To assert such control the new union turned to familiar practices reflective of its educational and political traditions.

In 1996 UNITE! built upon shocking public revelations of overseas sweatshops and child labor, in particular the disclosure that Kathy Lee Gifford's Wal-Mart clothing was produced under exploitative conditions. The union's antisweatshop campaign included picketing and protesting U.S. retailers known to carry apparel made by sweatshop labor as well as efforts to obtain representation on the board of a major U.S. retailer known to purchase "sweated" goods.[17]

UNITE! was also a major proponent, along with the National Consumers League, of the "Stop Sweatshops Bill," introduced in the U.S. Congress in October 1996. The bill proposed to amend the 1938 Fair Labor Standards Act to hold large manufacturers and retailers legally accountable for working conditions in domestic contractors' shops. The bill would make manufacturers and retailers liable for contractors' compliance with the provisions of the Fair Labor Standards Act regarding minimum wages and hours, child labor, and industrial homework.

In response to congressional inaction, during the last week of July 1998 several thousand union members from across the nation descended upon Washington to lobby Congress to support the legislation. UNITE! held a rally on the steps of the U.S. Capitol, then dispatched members to visit with legislators with the hope of securing more sponsors and bringing the measure to a committee vote. As of late summer 1998, the bill remained with the Committee on Education and the Workforce. Despite more than a hundred co-sponsors, the possibility of its enactment faded with the adjournment of

A modern U.S. garment factory in New York. (Courtesy of Kheel Center
Archives, Cornell University.)

the 105th Congress. Though reintroduced in the 106th Congress of 1999, it
was not enacted.[18]

UNITE! members also took part in massive protests at the third annual
ministerial conference of the World Trade Organization (WTO), held in
Seattle, Washington, in late 1999. In one of the largest and most violent
protests of its kind in recent times, scores of protestors disrupted trade
talks and demanded improved labor standards in developing economies
and a greater voice for workers in trade agreements. Environmental,
human, and labor rights issues were, once again, the central focus of those
concerned about negative repercussions of international trade policy.
Despite the perception of weakened clout and credibility, the WTO protests
defined the labor movement's militancy in the arena of international trade,
as did its opposition to the U.S. government's move to normalize trade
relations with the People's Republic of China in the spring of 2000.[19]

UNITE!'s position on China was emphasized when it challenged the Fair Labor Association's policies on monitoring of workplace conditions in the Asian nation. The voluntary codes of conduct and workplace inspection initiatives of the association, representing apparel makers, universities, and workers' and human rights organizations, are intended to mitigate labor abuse. UNITE! took a harder line and advocated that China and other countries ought to be off limits to the manufacture of garments bearing collegiate logos unless they dramatically changed their labor practices. When the association didn't agree, the union resigned its affiliation. In a related move, UNITE! has supported student-led campaigns to stop colleges and universities from selling apparel made with sweated labor. Student-inspired organizations, such as the United Students against Sweatshops and the Worker Rights Coalition, have been supported by the union.[20]

Postscript

In recent times there has been an upsurge of awareness across the United States of apparel sweatshops and the plight of the people who work in them, wherever they exist. This movement is apparent on college and university campuses. It is apparent in growing media coverage of raids on domestic sweatshops and on the practices of U.S. companies, such as Nike, that use overseas labor. Indeed, the mobilized opposition to the policies of the WTO and to "most favored nation" status for China attest to widespread awareness of labor exploitation. This renewed awareness is as much a human interest story in 2001 as it was when tragedy occurred at the Triangle Shirtwaist Company factory in 1911 and when the Uprising of 20,000 drew public attention in 1909. It is a contemporary awareness rooted in history.

History demonstrates that, with the help of many supporters, Min and Bill Matheson led a significant—and by most measures successful—effort in a remote area of Pennsylvania to enable garment workers to counter many of the same sweatshop conditions found today across the globe. Pennsylvania's Wyoming Valley District became one of the most well recognized and widely respected in the vast ILGWU network.

Jay Mazur, past president of UNITE!, addresses the Pennsylvania Labor History Society at the dedication of the historical marker erected to commemorate Min L. Matheson, Wilkes-Barre, September 1999. (Courtesy of George Zorgo, Zorgo Printing Service, Inc.)

In a new century, however, it is apparent that the struggle undertaken by the Mathesons is far from over. The battle has simply shifted across the vastness of the oceans and, in some instances, back to its points of origin in America's urban centers. Garment workers, both in the United States and in developing nations, face virtually the same problems as those of the immigrants on Manhattan's Lower East Side and of the wives and daughters of Pennsylvania's anthracite coal miners in the not too distant past. Indeed, the problems that plague the industry and its workers in the twenty-first century are remarkably similar to those in 1900.

Despite the efforts of the ILGWU and now UNITE!, a larger question remains: whether the labor movement has the wherewithal to counter the might and moderate the inequities of modern international capitalism. In these times of technological revolution, dramatic shifts in government policy, and ease of international capital movement, the economic barriers that once existed between nations are dissolving. We now live in an era when the financial affairs of nations are inextricably linked; when equity markets and investors determine capital investment; when corporations argue that remaining "competitive" hinges on cutting costs and enhancing stock value and shareholder dividends, often at the expense of workers; when short-term gains have priority over longer term contributions to communities, people, and families; and when developing nations—with the encouragement of the World Bank and the International Monetary Fund—embrace the free market as the panacea of "economic development."

Despite its well-meaning intentions, critical questions must be asked of the American labor movement. Do American unions like UNITE! possess the long-term internal viability, intellectual capability, financial means, and political capital to effectively influence these dramatic shifts in political economics? Does American labor sufficiently understand and take into consideration its own history and engage in active learning from past successes and mistakes? How can labor's intentions be effectively matched with action in domestic and, most important, international arenas? How is it possible for labor to promote humanitarian concerns, such as fair wages and safe working conditions, in an era when political scandals draw more press and public attention than joblessness, homelessness, underemployment, and poverty? All of these crucial questions find no easy answers.

The past contains some important lessons as to how such issues might be surmounted. Perhaps the ILGWU's work in Pennsylvania's anthracite coal fields in the latter two-thirds of the twentieth century can provide historically relevant insights from which perspectives and measures to address contemporary issues may be developed.

Appendix: Oral History Interviews

Unless we indicate otherwise, all interviews were conducted by one or more of the authors and are included in the Northeastern Pennsylvania Oral History Project, housed at the University of Wisconsin–Stevens Point. The Kheel Center for Labor Management Documentation and Archives, School of Industrial and Labor Relations, Cornell University, is cited as Kheel Center Archives. Not all of the interviews listed have been cited in the book, but each person provided important insights on the garment industry and ILGWU. The findings and conclusions expressed in this book are not necessarily those of the individuals who contributed oral histories. The date and location of each interview we conducted are indicated in parentheses at the end of the citation.

Berger, Martin, ILGWU assistant regional director of Northeast Department, staff member, organizer, and manager (Dec. 16, 1997, Harrisburg).

Bianco, Sam, ILGWU business agent, staff member, and Wyoming Valley District manager (Mar. 11, 1996, Wilkes-Barre).

Breslow, Israel, ILGWU vice president and manager of Local 22 (Mar. 2, 1982; Kheel Center Archives).

Caputo, Minnie, garment worker and ILGWU member (July 22, 1993, Pittston).

Cherkes, William, garment factory owner (July 20, 1994, Kingston, Pa.).

D'Angelo, Anthony, garment worker and barber (Dec. 17, 1988, and July 28, 1989, West Pittston).

DePasquale, Angelo, ILGWU organizer, Wyoming Valley District (July 23, 1993, Pittston Township).

Dubinsky, David, ILGWU president (June 25, July 10, and July 24, 1974; Kheel Center Archives).

Flood, Daniel, J. congressman, Eleventh District of Pennsylvania (July 5, 1990, Wilkes-Barre).

Gable, William, ILGWU staff member and chorus co-director and pianist (Dec. 7, 1999, Pittston).

Gingold, David, ILGWU Northeast Department vice president (July 25, 1974; Kheel Center Archives).

Greenberg, Betty, daughter of Min and Bill Matheson (Jan. 11, 1995, and Nov. 7, 1999, Kingston. Pa.).

Greenberg, Larry, son-in-law of Min and Bill Matheson (Nov. 7, 1999, Kingston, Pa.).

Gutstein, Leo, garment factory owner (June 26, 1997, Pittston).

Hartel, Lois, ILGWU Wyoming Valley district manager (June 17, 1993, Harrisburg, and Mar. 16, 1996, Wilkes-Barre).

Hoffman, Florice, Unity House employee and daughter of Sol Hoffman (Aug. 14, 1998, Unity House, Bushkill, Pa.).

Hoffman, Sol, ILGWU Northeast Department vice president and regional director (Mar. 6, 1998, New York; Aug. 14, 1998, Unity House, Bushkill, Pa.).

Hostetter, Robert, ILGWU education director, Northeast Department (July 3, 1993, Harrisburg).

Justin, John, ILGWU business agent and district manager in Harrisburg, Pottsville, and Easton, Pa., and Wilmington, Del. (Jan. 1, 1988; May 30, 1996; Nov. 25, 1997; and June 6, 2000, Wilkes-Barre).

Karp, Barnett, manager, ILGWU Philadelphia–South Jersey Joint Board (Jan. 5, 1976; interview by Marilyn Levin, his granddaughter).

Leader, George, governor of Pennsylvania, 1955–59 (May 30, 1995, Hershey, Pa.).

Levin, Edith, daughter of Barnett Karp (Oct. 23, 1998, Philadelphia).

Levin, Marilyn, ILGWU staff member and granddaughter of Barnett Karp (Oct. 23, 1998, Philadelphia).

Levin, Ted, Unity House employee and grandson of Barnett Karp (Oct. 23, 1998, Philadelphia).

Lieb, Ann, ILGWU staff member (June 30, 1997, Wilkes-Barre).

Lyons, Clementine, ILGWU business agent, education director, and chorus co-director (July 5, 1990, and July 24, 1993, Wyoming, Pa.).

Matheson, Min Lurye, ILGWU Wyoming Valley district manager (Nov. 20 and 21, 1982; Dec. 5, 1988; and June 28, 1990, by Robert Wolensky, Kingston; Sept. 7 and 21 and Oct. 27, 1983, and Sept. 23, 1985, by Alice Hoffman; Historical Collections and Labor Archives, Pennsylvania State University).

Matheson, Wilfred (Bill), ILGWU Wyoming Valley District education director (Mar. 24, 1981; Historical Collections and Labor Archives, Pennsylvania State University).

Mathews, Thomas, ILGWU assistant regional director, Northeast, Western Pennsylvania, and Ohio Department, educator, and manager (July 3, 1993, Harrisburg).

Mazur, Jay, past president of UNITE! (Mar. 6, 1998, New York).

Morand, Martin, ILGWU staff member, district manager, regional director Southeast Region, and professor (Dec. 4, 1997, Harrisburg).

Ney, Dorothy, ILGWU business agent (July 3, 1990, Wilkes-Barre).

Novak, Pearl, garment worker and ILGWU chorus member (June 19, 1997, and Oct. 11, 1998, Bear Creek, Pa.).

Reca, Alice, garment worker (July 26, 1995, Baltimore, Md.).

Reuter, Ralph, ILGWU director of health and welfare, Northeast Department (Mar. 6, 1998, New York; Aug. 14, 1998, Unity House, Bushkill, Pa.).

Sampiero, Helen, sister of Min Matheson (Mar. 22, 1995, Kingston, Pa.).

Silverman, Jennie, garment worker and ILGWU member (July 16, 1993, New York).

Solomon, Irwin, ILGWU secretary-treasurer (Mar. 5, 1998, New York).

Weiss, Helen, ILGWU business agent (June 20, 1997, Butler, N.J.).

Whittaker, Celina, Unity House employee (Apr. 19, 1999, Bushkill, Pa.).

Whittaker, Nelson, Sr., Unity House employee (Apr. 19, 1999, Bushkill, Pa.).

Whittaker, Nelson, Jr., Unity House employee (Aug. 14, 1998, Bushkill, Pa.).

Williams, Joseph, member of Wilkes-Barre City Council (July 31, 1984, Wilkes-Barre).

Zimmerman, Charles (Sasha), ILGWU vice president, manager of New York

Dress Joint Board and Local 22 (May 16, 1977, and Jan. 5, 1978; Kheel Center Archives).

Zobel, Sonia, vacationer at Unity House (Mar. 5, 1998, New York).

Zorgo, George, Sr., printing company owner, friend of Min and Bill Matheson (July 26, 1993, West Pittston).

Notes

Introduction

1. On deindustrialization, see, for example, B. Bluestone and B. Harrison, *The Deindustrialization of America* (New York: Basic Books, 1992). On deindustrialization in Pennsylvania see, for example, M. Descht, "Work: With Selected Photographs from the Exhibit 'Faces from an American Dream,'" *Pennsylvania History* 65 (Summer 1998): 368–81; "Deindustrialization: A Panel Discussion," *Pennsylvania History* 58 (July 1991): 118–21; T. Dublin, *When the Mines Closed* (Ithaca, N.Y.: Cornell University Press, 1998); J. Hoerr, *And the Wolf Finally Came* (Pittsburgh: University of Pittsburgh Press, 1988); D. Miller and R. Sharpless, *The Kingdom of Coal* (Philadelphia: University of Pennsylvania Press, 1985); and R. Wolensky, K. Wolensky, and N. Wolensky, *The Knox Mine Disaster: The Final Years of the Northern Anthracite Industry and the Effort to Rebuild a Regional Economy* (Harrisburg: Pennsylvania Historical Museum Commission, 1999).

2. A few publications focus specifically on the history of the apparel industry and the ILGWU or the Amalgamated Clothing and Textile Workers in areas other than New York. See, for example, A. Bisno, *Abraham Bisno, Union Pioneer* (Madison: University of Wisconsin Press, 1967); W. Carsel, *A History of the Chicago Ladies' Garment Workers' Union* (Chicago: Normandie House, 1940); J. Laslet and M. Tyler, *The ILGWU in Los Angeles, 1907–1988* (Los Angeles: Ten Star Press, 1989); E. LeMar, *The Clothing Workers in Philadelphia* (Philadelphia: Joint Board of the Amalgamated Clothing Workers of America, 1940); R. Schneiderman and L. Goldthwaite, *All for One* (New York: Paul S. Erickson, 1967); K. Wolensky and R. Wolensky, "Born to Organize," *Pennsylvania Heritage* 25, no. 3 (Summer 1999): 32–39; and R. Wolensky and K. Wolensky, "Min Matheson and the ILGWU in the Northern Anthracite Region," *Pennsylvania History* 60, no. 4 (October 1993): 455–74. Also, ILGWU, *News History: The Story of the Ladies' Garment Workers* (New York, 1950), includes excerpts and discussion of organizing and union activities throughout the United States and Canada.

3. S. Hoffman, oral history interview (joint interview with R. Reuter), Mar. 6, 1998, Northeastern Pennsylvania Oral History Project, University of Wisconsin–Stevens Point (hereafter NPOHP), tape 1, side 2.

4. Min Matheson was interviewed by Robert Wolensky on four occasions: Nov. 20 and 21, 1982; Dec. 5, 1988; and June 28, 1990. All interviews and transcripts are available as part of the Northeastern Pennsylvania Oral History Project, housed at the Center for the Small City, University of Wisconsin–Stevens Point. Although Bill Matheson was present at the first two of these interviews, Alzheimer's disease prevented him from contributing to the narratives.

Notes

Throughout the text all direct quotes from Min Matheson are cited from this collection. References to her interviews with Alice Hoffman (see note 6) in which she provided substantiating or similar information are provided in the notes to the text.

5. Because oral history offers "reminiscences, accounts, and interpretations of events from the recent past" (see A. Hoffman, "Reliability and Validity in Oral History," in *Oral History: An Interdisciplinary Anthology*, ed. D. Dunaway and W. Baum [Nashville: American Association for State and Local History, 1984], 68), we believe that it provides one of the most compelling methods available to document the historical record of the ILGWU in Pennsylvania's Wyoming Valley. However, while this study relies a great deal on oral history, we have used primary and secondary sources extensively as well. These sources complement one another to tell a more complete story. Sources on oral history as a research and interpretive tool are many. In addition to the Hoffman article cited above, relevant works include J. Bennett, "Human Values in Oral History," *Oral History Review* 11 (Fall 1983): 1–15; M. Frisch, *A Shared Authority* (Albany: SUNY Press, 1990); R. Grele, *Envelopes of Sound* (New York: Praeger, 1991); K. Olson and L. Shopes, "Crossing Boundaries, Building Bridges: Doing Oral History among Working Class Women and Men," in *Women's Words: The Feminist Practice of Oral History*, ed. S. Gluck and D. Patai, 189–204 (New York: Routledge, 1991); C. Oblinger, *Interviewing the People of Pennsylvania* (Harrisburg: Pennsylvania Historical and Museum Commission, 1978); and A. Portelli, *The Death of Luigi Trastulli and Other Stories* (Albany: SUNY Press, 1991). Finally, for an excellent source on oral history research in Pennsylvania, see *Pennsylvania History* 60 (October 1993), special issue: *Oral History in Pennsylvania*, ed. L. Shopes.This issue features R. Wolensky and K. Wolensky, "Min Matheson and the ILGWU in the Northern Anthracite Region," 477–74.

6. Min Matheson was interviewed by Alice Hoffman on four occasions: Sept. 7, Sept. 21, and Oct. 27, 1983, and Sept. 23, 1985. Lois Hartel, who later served in Min's capacity with the ILGWU in the northern anthracite region , was present at these interviews. Bill Matheson was also interviewed by Alice Hoffman on Mar. 24, 1981. His interview focused more on workers' education and activism than on the ILGWU's experiences in Pennsylvania. These interviews and transcriptions are available at the Historical Collections and Labor Archives, Paterno Library, Pennsylvania State University, University Park.

7. M. Kaufmann, personal communication, Mar. 28, 1997.

1. An Industry, a Union, and Runaway Garment Factories

1. Jewish immigration from 1882 to 1900, for example, totaled nearly 600,000 persons.

2. Several sources discuss the evolution and history of the women's garment industry and unionization in the United States in the nineteenth and early twentieth centuries. See Melech Epstein, *Jewish Labor in the U.S.A.* (Hoboken, N.J.: KTAV, 1969); L. Lorwin, *The Women's Garment Workers: A History of the International Ladies' Garment Workers' Union* (New York: Arno, 1969), esp. chaps. 1–6; D. Montgomery, *The Fall of the House of Labor* (New York: Cambridge University Press, 1987), esp. 116–23; L. Stein, ed., *Out of the Sweatshop: The Struggle for Industrial Democracy* (New York: Quadrangle, 1977); L. Teper, *The Women's Garment Industry: An Economic Analysis* (New York: ILGWU Educational Department, 1937); G. Tyler, *Look for the Union Label: A History of the ILGWU* (Armonk, N.Y.: M. E. Sharpe, 1995); and R. Waldinger, *Through the Eye of the Needle* (New York: New York University Press, 1986). For discussion of the role of immigrants and women in industry, including apparel making, see L. Dinnerstein

and D. Reimers, *Ethnic Americans: A History of Immigration and Assimilation* (New York: Harper & Row, 1975); S. Glenn, *Daughters of the Shtetl: Life and Labor in the Immigrant Generation* (Ithaca, N.Y.: Cornell University Press, 1991); and R. Takaki, *A Different Mirror: A History of Multiculturalism in America* (Boston: Little, Brown, 1993).

3. On ILGWU history, see M. Danish, *The World of David Dubinsky* (Cleveland: World, 1957); D. Dubinsky, *A Life with Labor* (New York: Simon & Schuster, 1977); Epstein, *Jewish Labor in the U.S.A.*; P. Foner, *History of the Labor Movement in the United States*, vol. 9 (New York: International Publishers, 1991); ILGWU, *News History: The Story of the Ladies' Garment Workers* (New York, 1950); Lorwin, *Women's Garment Workers*; Stein, *Out of the Sweatshop*; B. Stolberg, *Tailor's Progress: The Story of a Famous Union and the Men Who Made It* (New York: Doran, Doubleday, 1944); and Tyler, *Look for the Union Label*.

4. Historically apparel has been produced by three types of firms. The manufacturer designs the garments, purchases the fabric, maintains a factory where workers produce garments, and sells the finished products to a retailer. A jobber does all of the tasks of a manufacturer with one exception: jobbers typically do not own factories or employ workers; rather, they turn to contractors to produce the finished products. Jobbers can—and often do—have numerous arrangements with contractors to produce finished goods. A contractor sews the cut material into garments by contract with a jobber or manufacturer. The task of the contractor is to produce the finished product according to the specifications of the manufacturer or jobber.

5. For a firsthand account of the Uprising of 20,000, see T. Malkiel, *The Diary of a Shirtwaist Striker* (Ithaca, N.Y.: Cornell University Press, 1990), which recounts the events of 1909 from a diary published by Malkiel in 1910. For other accounts of the Uprising, the Great Revolt, and the Triangle fire, see ILGWU, *News History*; R. Schneiderman and L. Goldthwaite, *All for One* (New York: Paul F. Erickson, 1967); Stein, *Out of the Sweatshop*, esp. chaps. 4, 5, and 8; L. Stein, *The Triangle Fire* (New York: Carroll, Graf & Quicksilver Books, 1962); Stolberg, *Tailor's Progress*; and Tyler, *Look for the Union Label*, esp. chaps. 5, 6, and 7.

6. For discussion of the contracting system, see ILGWU, *News History*; Lorwin, *Women's Garment Workers*, esp. chap. 3; Stolberg, *Tailor's Progress*; Teper, *Women's Garment Industry*, esp. 6–8; Tyler, *Look for the Union Label*. Also helpful on this topic is an unpublished manuscript prepared in 1951 by the ILGWU attorney Emil Schlesinger titled "The Outside System of Production in the Women's Garment Industry in New York." A copy of the manuscript was provided by the ILGWU (now UNITE!) research department.

7. On early twentieth-century organizing work of the ILGWU in Philadelphia, see the following reports of the ILGWU: *Report and Proceedings of the Tenth Convention of the ILGWU*, June 6–11, 1910 (New York, 1910); *Report and Proceedings of the Eleventh Convention of the ILGWU*, June 3–12, 1912 (New York, 1912); *Report and Proceedings of the Twelfth Convention of the ILGWU*, June 1–13, 1914 (New York, 1914); *Report and Proceedings of the Thirteenth Convention of the ILGWU*, Oct. 16–28, 1916 (New York, 1916); *Report and Proceedings of the Fourteenth Convention of the ILGWU*, May 20–June 1, 1918 (New York, 1918); *Report and Proceedings of the Fifteenth Convention of the ILGWU*, May 3–15, 1920 (New York, 1920).

8. Tyler, *Look for the Union Label*, 157.

9. The term "joint board," as used by the ILGWU, dates to its formation in 1900. A joint board is an entity representing workers from each of the trades associated with a specific division of the industry or in a specific geographic region. Each local union, representing a trade, elects members to the Joint Board. The board negotiates agreements with employers, enforces standards of wages, hours, and working conditions, organizes shops, and engages in other

242

Notes

related activities. The New York Cloak Joint Board, for example, traditionally included representatives of each of the local unions in the cloakmaking sector of the industry, such as Local 117, Cloak Operators; Local 9, Cloak Finishers; Local 35, Cloak Pressers; and Local 10, Cutters. The New York Dress Joint Board was organized along similar craft lines. The Philadelphia Joint Board was organized along geographic lines.

10. On Communist influence in the ILGWU in the 1920s and the near destruction of the union as a result, see Epstein, *Jewish Labor in the U.S.A.*, esp. 124–56; and the following reports of the ILGWU: *Report and Proceedings of the Seventeenth Convention of the ILGWU,* May 5–17, 1924 (New York, 1924); *Report and Proceedings of the Eighteenth Convention of the ILGWU,* Nov. 30–Dec. 17, 1925 (New York, 1925); and *Report and Proceedings of the Nineteenth Convention of the ILGWU,* May 5–17, 1928 (New York, 1928). Also see J. Barrett, "Boring from Within and Without: W. Z. Foster, the Trade Union Educational League, and American Communism in the 1920s," in *Labor Histories: Class, Politics, and Working-Class Experience,* ed. E. Arensen, J. Greene, and B. Laurie (Urbana: University of Illinois Press, 1998), 309–39; ILGWU, *News History;* Stolberg, *Tailor's Progress;* and Tyler, *Look for the Union Label,* esp. chap. 12. Initially the Communist influence was greatest in Local 25. In an attempt to isolate this influence in 1921, the ILGWU leadership split the dressmakers from Local 25 and created a separate dressmakers' union known as Local 22.

11. J. Silverman, oral history interview, July 16, 1993, Northeastern Pennsylvania Oral History Project, University of Wisconsin–Stevens Point (hereafter NPOHP), tape 1, side 1.

12. Schlesinger and Dubinsky shared the view that the growing dress industry required a separate union joint board. Thus they reversed Sigman's earlier decision to establish a single joint board and recreated a cloak joint board and a dress joint board.

13. ILGWU, *Report of the General Executive Board to the Sixteenth Convention of the ILGWU,* May 1, 1922 (New York, 1922), and *Report and Proceedings of the Eighteenth Convention of the ILGWU,* Nov. 30–Dec. 17, 1925 (New York, 1925).

14. A 1933 contract between the cloak and suit industry and the New York Cloak Joint Board held jobbers accountable for wages paid in contract shops, restricted the ability of a jobber to change contractors, and enabled the ILGWU to negotiate piece rates with jobbers and enforce such rates in contract shops. These provisions were extended to the dress industry in a 1936 agreement with the New York Dress Joint Board. As Chapter 6 shows, however, such provisions became nearly impossible to enforce as nonunion contract shops sprouted in remote areas.

15. ILGWU, *Report of the General Executive Board to the Twenty-first Convention of the ILGWU,* May 2, 1932 (New York, 1932), 14.

16. S. Chaikin, *A Labor Viewpoint: Another Opinion* (Monroe, N.Y.: Library Research Associates, 1980), 110–11.

17. F. Perkins (1933), "The Cost of a Five-Dollar Dress," in Stein, *Out of the Sweatshop,* 224–25.

18. *Christian Science Monitor,* "The N.R.A. Codes," in Stein, *Out of the Sweatshop,* 229–30.

19. E. Butler, *Women and the Trades: Pittsburgh, 1907–1908* (New York: Charities Publication Committee, 1909), 130. On working conditions in Pennsylvania apparel and textile factories in the late nineteenth and early twentieth centuries and the enactment of the 1895 Sweatshop Law, see the following reports of the Commonwealth of Pennsylvania: *Sixth Annual Report of the Factory Inspector* (Harrisburg, 1896); *Tenth Annual Report of the Factory Inspector* (Harrisburg, 1901); and *Twelfth Annual Report of the Factory Inspector* (Harrisburg, 1903).

20. Commonwealth of Pennsylvania, *Message of Governor Gifford Pinchot to the General*

Assembly in Joint Session, January 4, 1927 (Harrisburg, 1927). On child and women's labor and related legislation in Pennsylvania, see P. Klein and A. Hogenboom, *A History of Pennsylvania* (University Park: Penn State University Press, 1973); H. Harris and P. Blatz, eds., *Keystone of Democracy: A History of Pennsylvania Workers* (Harrisburg: Pennsylvania Historical and Museum Commission, 1999); and K. Wolensky and J. Rich, *Child Labor in Pennsylvania*, Historic Pennsylvania Leaflet no. 46 (Harrisburg: Pennsylvania Historical and Museum Commission, 1998).

21. Commonwealth of Pennsylvania, *Pennsylvania Labor and Industry in the Depression*, Special Bulletin no. 39 (Harrisburg: Department of Labor and Industry, 1934).

22. Commonwealth of Pennsylvania, *Message of Governor Gifford Pinchot to the General Assembly in Joint Session, January 4, 1935* (Harrisburg, 1935). For a discussion of the life of Cornelia Bryce Pinchot, see M. Voda, "The Lady in Red: Cornelia Bryce Pinchot—Feminist for Social Justice," *Pennsylvania Heritage* 23, no. 4 (Fall 1997): 22–31.

23. D. Cupper, *Working in Pennsylvania: A History of the Department of Labor and Industry* (Harrisburg: Pennsylvania Historical and Museum Commission, 2000); on the Department of Labor and Industry during the Great Depression, see chap. 2, esp. 36–37.

24. For further discussion of the anthracite industry and ethnicity and immigration in the anthracite region, see H. Aurand, *From the Mollie Maguires to the United Mine Workers: The Social Ecology of an Industrial Union* (Philadelphia: Temple University Press, 1971); J. Bodnar, *Anthracite People: Families, Work, and Unions, 1900–1940* (Harrisburg: Pennsylvania Historical and Museum Commission, 1983); E. J. Davies, *The Anthracite Aristocracy: Leadership and Social Change in the Hard Coal Regions of Northeastern Pennsylvania , 1800–1930* (De Kalb: Northern Illinois University Press, 1985); T. Dublin, *When the Mines Closed* (Ithaca, N.Y.: Cornell University Press, 1998); D. Miller and R. Sharpless, *The Kingdom of Coal: Work, Enterprise, and Ethnic Communities in the Mine Fields* (Philadelphia: University of Pennsylvania Press, 1985); E. W. Roberts, *The Breaker Whistle Blows: Mining Disasters and Labor Leaders in the Anthracite Region* (Scranton: Anthracite Museum Press, 1984); Z. Wasyliw, "European Identities of East Slavic Settlements in Northeastern Pennsylvania during the Late Nineteenth and Early Twentieth Centuries," in *Proceedings of the Sixth Annual Conference on the History of Northeastern Pennsylvania*, ed. R. Janosov, 116–32 (Nanticoke, Pa.: Luzerne County Community College, 1994); A. F. C. Wallace, *St. Clair: A Nineteenth-Century Coal Town's Experience with a Disaster-Prone Industry* (New York: Knopf, 1987); and R. Wolensky, K. Wolensky, and N. Wolensky, *Final Breach: The Knox Mine Disaster, the Contract-Leasing System, and the Demise of the Northern Anthracite Industry* (Urbana: University of Illinois Press, forthcoming).

25. On the Knox Mine Disaster, see R. Wolensky and K. Wolensky, "Disaster—or Murder?—in the Mines," *Pennsylvania Heritage* 24, no. 3 (Spring 1998): 23–28; R. Wolensky, K. Wolensky, and N. Wolensky, *The Knox Mine Disaster: The Final Years of the Northern Anthracite Industry and the Effort to Rebuild a Regional Economy* (Harrisburg: Pennsylvania Historical and Museum Commission, 1999); and Wolensky et al., *Final Breach*.

26. ILGWU, *Report of the General Executive Board to the Twenty-second Convention of the ILGWU*, May 28, 1934 (New York: ABCO Press, 1934), 34.

27. ILGWU, Philadelphia Joint Board, minutes of Executive Board meetings, Dec. 21, 1937; Mar. 27, 1939; and Apr. 25, 1939, Urban Archives, Temple University, ILGWU Executive Board Minutes of Meetings, 1937–58, no. 544, box 1, folder 1. ILGWU organizing in Philadelphia during this period was described by Barnett Karp, oral history interview, January 1976, provided by Marilyn Levin, Karp's granddaughter, who conducted the interview.

28. W. Cherkes, oral history interview, July 20, 1994, NPOHP, tape 1, side 1.

29. Ibid.

30. L. Gutstein, oral history interview, July 26, 1997, NPOHP, tape 1, side 1.

31. J. Justin, oral history interview, Jan. 28, 1988, NPOHP, tape 1, side 1.

32. Cherkes oral history interview, tape 1, side 1.

33. K. Munley, "Elizabeth R. Lynett: Crusader for Women's Rights," paper presented at the symposium "In Celebration of Pennsylvania's Women in the Labor Movement," Marywood University, Scranton, Nov. 15, 2000. For examples of Lynett's newspaper articles, see "Learners Preferred to Experienced Hands in Mills," *Scranton Times*, May 26, 1933, 3; "Misery for Thousands Price Paid for 'Bargains,'" *Scranton Times*, May 29, 1933, 3, 9.

34. Pennsylvania Crime Commission, *Report on Organized Crime* (Harrisburg, 1970), 8. For a discussion of organized crime in New York's garment industry, see Tyler, *Look for the Union Label*, esp. chap. 18. An important source on the history of organized crime is S. Fox, *Blood and Power: Organized Crime in Twentieth-Century America* (New York: Penguin, 1989).

35. L. Gutstein oral history interview, tape 1, side 1.

36. Pennsylvania Power and Light (PP&L) data on the number of garment factories in the Wyoming Valley were provided by Thomas Dublin of the State University of New York at Binghamton. Using PP&L reports and quantitative information, Dublin has created an SPSS database to study anthracite region economic revitalization efforts and the impact of the Pennsylvania Industrial Development Authority (discussed in Chapter 5). Data on the number of garment factories in Pittston are derived from an undated memo headed "Members of the Greater Pittston Contractors Association," David Dubinsky Papers, 5780/002, box 297, folder 4A, Kheel Center for Labor Management Documentation and Archives, School of Industrial and Labor Relations, Cornell University. Cornell University's School of Industrial and Labor Relations is the official archive of the ILGWU. For obvious reasons, it is very difficult to determine the exact number of factories owned, controlled, or influenced by members of organized crime. According to Clementine Lyons and Dorothy Ney, both of whom worked closely with Min Matheson, while actual syndicate ownership was probably relatively small—no more than fifteen to twenty shops throughout the entire valley—some garment factories were influenced by criminal elements, as legitimate owners feared retaliation if they appeared friendly to the ILGWU. The situation was unquestionably more apparent in Pittston. See C. Lyons, oral history interview, July 5, 1990, NPOHP, tape 1, side 1, and D. Ney, oral history interview, June 26, 1993, NPOHP, tape 1, side 1.

37. L. Teper, *Women's Garment Industry;* ILGWU, *Report of the General Executive Board to the 22nd Convention of the ILGWU*, May 3, 1937 (New York: ABCO Press, 1937).

38. Wage differences between New York and Pennsylvania, discussed in more detail in Chapter 6, were a major cause of the 1958 general dress industry strike.

39. The role of the Dress Joint Board was to administer contracts with the industry. It was made up of four New York City locals of the ILGWU: Local 89 (Italian sewing machine operators and pressers), Local 22 (Jewish sewing machine operators), Local 10 (fabric/material cutters), and Local 60 (Jewish pressers). The locals were established around the ethnic and craft identities of New York garment workers and provided important membership services while facilitating communication with and between non-English-speaking members. A dress factory often included members of each local.

40. D. Melman, "The Cause and Effect of the ILGWU Dress Industry General Strike of 1958," master's thesis, Baruch College, City University of New York, and School of Industrial and Labor Relations, Cornell University, 1994, 11–15.

41. "Little David, the Giant," *Time*, Aug. 29, 1949, 12–15.

Notes

2. The ILGWU's Response to the Runaways

1. Gingold recounts his experiences in Pennsylvania and in the anthracite region in H. Crone, 35 Northeast: A Short History of the Northeast Department of the ILGWU Based upon the Reminiscences and Diaries of David Gingold and Official ILGWU Records: 1935–1970 (New York: ILGWU, 1970).

2. On David Gingold's life and career with the ILGWU, see P. Seegal, Double Texture: A Profile of David Gingold (New York: ILGWU, 1953).

3. Justice, Oct. 15, 1936, 4.

4. D. Gingold, oral history interview, July 25, 1974, Kheel Center for Labor Management Documentation and Archives, School of Industrial and Labor Relations, Cornell University (hereafter Kheel Center Archives), Oral History no. 3, 68–85.

5. Seegal, Double Texture, 15.

6. ILGWU, Report of the General Executive Board to the 22nd Convention of the ILGWU, May 3, 1937 (New York: ABCO Press, 1937).

7. Gingold, oral history interview, 83–86.

8. ILGWU, Report of the General Executive Board to the 22nd Convention.

9. ILGWU, Report of the General Executive Board to the 24th Convention of the ILGWU, May 27–June 8, 1940 (New York: ABCO Press, 1940).

10. Crone, 35 Northeast, 60; ILGWU, Report of the General Executive Board to the 25th Convention of the ILGWU, May 29–June 9, 1944 (New York: ABCO Press, 1944).

11. The 1940 Report of the General Executive Board of the ILGWU to the 24th Convention reported that the union "is now in the process of organizing one of the largest knitted underwear firms in the country—Belle Knitting Mills of Sayre, Pa., employing over 1,000 workers" (74). The union succeeded in organizing Belle in 1941 and formed Local 365 in Sayre. Bill Matheson worked for the local. Also see Sayre Historical Society, Sayre Quarterly 7, no. 4 (Winter 1995–96).

12. Min recalled that the ILGWU organizer William Ross initially approached her and Bill about relocating to the Wyoming Valley in late 1943 or early 1944, and that Dubinsky then contacted them to confirm the request. Ross later became manager of the Philadelphia–South Jersey Joint Board of the ILGWU. See M. Matheson, oral history interview, Nov. 30, 1982, Northeastern Pennsylvania Oral History Project, University of Wisconsin–Stevens Point (hereafter NPOHP), tape 1, side 1. Also see M. Matheson, oral history interview, Sept. 7, 1983, Oral History Collection, Historical Collections and Labor Archives, Pennsylvania State University, University Park (hereafter HCLA), tape 1, side 1.

13. M. Matheson, oral history interview, Nov. 30, 1982, tape 1, side 1; see also her oral history interview, Sept. 7, 1983, tape 1, side 1.

14. M. Matheson, oral history interview, Nov. 30, 1982, tape 1, side 1.

15. M. Matheson, oral history interviews, Dec. 5, 1988, NPOHP, tape 1, side 1, and Sept. 7, 1983, tape 1, side 2.

16. M. Matheson, oral history interview, Nov. 30, 1982, tape 1, side 1.

17. J. Silverman, oral history interview, July 16, 1993, NPOHP, tape 1, side 2. Additional information on Min's early years with the ILGWU and Local 22 in New York was obtained from an undated ILGWU press release announcing her appointment as director of the Union Label Office, David Dubinsky Papers, Kheel Center Archives, box 281, folder 5.

18. I. Breslow, oral history interview, Mar. 2, 1982, Kheel Center Archives, Oral History no. 24, box 19, vol. 3, 392–407.

19. M. Matheson, oral history interview, Dec. 5, 1988, tape 1, side 2.

20. B. Greenberg, oral history interview, Jan. 11, 1995, NPOHP, tape 1, side 1.

21. M. Matheson, oral history interviews, Dec. 5, 1988, tape 1, side 2, and Oct. 27, 1983, HCLA, tape 1, sides 1 and 2.

22. Greenberg, oral history interview, Jan. 11, 1995, tape 1, side 1.

23. M. Matheson, oral history interview, Dec. 5, 1988, tape 1, side 2.

24. M. Caputo, oral history interview, July 22, 1993, NPOHP, tape 1, side 1.

25. Christian Science Monitor, "The N.R.A. Codes," in Out of the Sweatshop: The Struggle for Industrial Democracy, ed. L. Stein (New York: Quadrangle, 1977), 229–30.

26. M. Matheson, oral history interview, Dec. 5, 1988, tape 1, side 2.

27. M. Matheson, oral history interviews, Nov. 30, 1982, tape 2, side 1, and Dec. 5, 1988, tape 1, side 2. Min discusses Pittston, in particular, as a "controlled town" in her oral history interview of Oct. 27, 1983, tape 1, side 2.

28. M. Matheson, oral history interviews, Dec. 5, 1988, tape 1, side 2, and Oct. 27, 1983, tape 1, side 1.

29. P. Jacobs, "David Dubinsky: Why His Throne Is Wobbling," in Labor: Readings on Major Issues, ed. R. Lester, 105–18 (New York: Random House, 1962).

30. L. Gutstein, oral history interview, July 26, 1997, NPOHP, tape 1, side 1.

3. Strategizing and Organizing

1. M. Matheson, oral history interview, Nov. 30, 1982, Northeastern Pennsylvania Oral History Project, University of Wisconsin–Stevens Point (hereafter NPOHP), tape 2, side 1.

2. J. Silverman, oral history interview, July 16, 1993, NPOHP, tape 2, side 1.

3. U.S. Senate Permanent Subcommittee on Investigations, Committee on Government Operations, Organized Crime and Illicit Traffic in Narcotics, 88th Cong., 1st sess., 1963, S. Rept. 72, 386. See also Pennsylvania Crime Commission, Report on Organized Crime (Harrisburg, 1970) and A Decade of Organized Crime (Harrisburg, 1980); S. Fox, Blood and Power: Organized Crime in Twentieth-Century America (New York: Penguin, 1989) esp. 326–29, 337.

4. M. Matheson, oral history interview, Nov. 30, 1982, NPOHP, tape 2, side 1. See also her oral history interview, Oct. 27, 1983, Oral History Collection, Historical Collections and Labor Archives, Pennsylvania State University (hereafter HCLA), tape 1, side 2.

5. A. D'Angelo, oral history interview, Dec. 17, 1988, NPOHP, tape 1, side 1; July 28, 1989, NPOHP, tape 1, side 2.

6. J. Justin, oral history interview, Jan. 28, 1988, NPOHP, tape 1, side 1.

7. B. Greenberg, oral history interview, Jan. 11, 1995, NPOHP, tape 1, side 1.

8. Several sources discuss the UMWA in the anthracite region. See, for example, H. Aurand, From the Mollie Maguires to the United Mine Workers: The Social Ecology of an Industrial Union (Philadelphia: Temple University Press, 1971); J. Bodnar, Anthracite People: Families, Work, and Unions, 1900–1940 (Harrisburg: Pennsylvania Historical and Museum Commission, 1983); P. Blatz, Democratic Miners: Work and Labor Relations in the Anthracite Coal Industry, 1875–1925 (Albany: SUNY Press, 1994); H. Harris and P. Blatz, eds., Keystone of Democracy: A History of Pennsylvania Workers (Harrisburg: Pennsylvania Historical and Museum Commission, 1999); D. Miller and R. Sharpless, The Kingdom of Coal: Work, Enterprise, and Ethnic Communities in the Mine Fields (Philadelphia: University of Pennsylvania Press, 1985); D. Monroe, "A Decade of Turmoil: John L. Lewis and the Anthracite Miners," Ph.D. dissertation, George-

town University, 1977; E. Roberts, *The Breaker Whistle Blows: Mining Disasters and Labor Leaders in the Anthracite Region* (Scranton: Anthracite Museum Press, 1984); and R. Wolensky, K. Wolensky, and N. Wolensky, *The Knox Mine Disaster : The Final Years of the Northern Anthracite Industry and the Effort to Rebuild a Regional Economy* (Harrisburg: Pennsylvania Historical and Museum Commission, 1999), and *Final Breach: The Knox Mine Disaster, the Contract-Leasing System, and the Demise of the Northern Anthracite Industry* (Urbana: University of Illinois Press, forthcoming).

9. M. Matheson, oral history interview, Nov. 30, 1982, tape 2, side 1.

10. J. Justin, oral history interview, Jan. 28, 1988, tape 1, side 1.

11. A. DePasquale, oral history interview, July 23, 1993, NPOHP, tape 1, side 1.

12. Ibid.

13. M. Matheson, oral history interview, Nov. 30, 1982, tape 2, side 2.

14. A. DePasquale, oral history interview, July 23, 1993, tape 1, side 1.

15. M. Matheson, oral history interview, Nov. 30, 1982, tape 2, side 2.

16. D. Ney, oral history interview, July 3, 1990, NPOHP, tape 1, side 1.

17. M. Matheson, oral history interview, Nov. 30, 1982, tape 2, side 2. On the Northeast Department's efforts to organize McKitterick-Williams, see ILGWU, *Report of the General Executive Board to the Twenty-sixth Convention of the ILGWU,* June 16–26, 1947 (New York: ABCO Press, 1947), esp. 116.

18. M. Matheson, oral history interviews, Nov. 30, 1982, tape 1, side 1; Dec. 5, 1988, NPOHP, tape 1, side 1; Oct. 27, 1983, tape 2, side 1.

19. M. Matheson, oral history interview, Dec. 5, 1988, tape 1, side 1. On the Committee of 100, see S. Spear, *Wyoming Valley History Revisited* (Shavertown, Pa.: Jemags, 1994), esp. chap. 20.

20. ILGWU, *Report of the General Executive Board to the Twenty-sixth Convention.*

21. M. Matheson, oral history interview, Dec. 5, 1988, tape 1, side 2.

22. Ibid. For further discussion of these and other stories relating to Min Matheson's tactics in countering organized crime in Pittston, see "The Lady and the Gangsters," *Justice,* March 1993; and L. Veely, "Min Matheson's War against Pittston's Gangsters," *Reader's Digest,* January 1957.

23. B. Greenberg, oral history interview, Jan. 11, 1995, tape 1, side 2.

24. A. DePasquale, oral history interview, July 23, 1993, tape 1, side 2.

25. Ibid.

26. M. Caputo, oral history interview, July 22, 1993, NPOHP, tape 1, side 1.

27. J. Silverman, oral history interview, July 16, 1993, NPOHP, tape 2, side 2.

28. B. Greenberg, oral history interview, Jan. 11, 1995, tape 1, side 2.

29. ILGWU, *Report of the General Executive Board to the Twenty-seventh Convention of the ILGWU,* May 23–June 1, 1950 (New York: ABCO Press, 1950).

30. Pennsylvania Department of Labor and Industry, Labor Dispute Case Files—Clothing Industry, Pennsylvania State Archives, RG 16, box 3, folder 1, 1951.

31. Ibid.

32. "Wilkes-Barre Mop-Up Annexes 4 Small Firms, 150 Members," *Justice,* Feb. 1, 1953, 9; "Organizing Brings 1,200 to Wilkes-Barre Roster," *Justice,* Apr. 15, 1953, 9.

33. "Emkay Talks Stall as Firm Balks at Union Demands," *Justice,* Mar. 15, 1953, 9; "Workers Vote 4 to 1 for ILGWU," *Justice,* Oct. 15, 1953, 3. Settlement terms of the Emkay contract are detailed in a memo in David Dubinsky Papers, Kheel Center for Labor Management Documentation and Archives, School of Industrial and Labor Relations, Cornell University (hereafter Kheel Center Archives), 5780/002, box 319, folder 12C.

34. "New Pacts at Pennsy Shops Win Hours Slash for 2,000," *Justice*, Feb. 15, 1954, 10.
35. "Northeast Enrolls 5 Pennsy Companies—300 Win ILG Benefits," *Justice*, Feb. 1, 1955, 2; "Top Court Upholds Right of ILG to Picket Peacefully at Wilkes Firm," ibid., 9.
36. "Pa. Rules ILG Welfare Pay No Bar to Jobless Benefits," *Justice*, Jan. 15, 1955, 2.
37. ILGWU, *Report of the General Executive Board to the Twenty-eighth Convention of the ILGWU*, May 18, 1953 (New York: ABCO Press, 1953).
38. B. Bliven, "Wm. Lurye Dies Fighting Open Shops," *New Republic*, June 20, 1949, 10; I. Lieberman and R. Williams, "Thousands Mourn Slain Organizer: Union Promises to Avenge Martyr," *New York Post*, May 12, 1949, 2.
39. David Dubinsky to Mrs. Wm. Lurye, May 9, 1949, telegram, Dubinsky Papers, box 319, folder 12B.
40. "Statement of Wyoming Valley and Greater Pittston District Councils of the ILGWU on the Murder of William Lurye," May 17, 1949, ibid.
41. Personal correspondence and memos on the murder of William Lurye, May 1949, ibid. Instructions to New York garment district workers on their work stoppage and participation in the Lurye funeral procession are detailed in a memo headed "Line of March," Charles Zimmerman Papers, 5780/014, Kheel Center Archives, box 24, folder 5.
42. Betty Greenberg discusses the murder of Will Lurye and its impact on her mother in an oral history interview, Jan. 11, 1995, tape 1, side 2.
43. "Remember Will Lurye," *New York Post*, May 9, 1950, 6. Background information on the police investigation of Macri and Guisto and the subsequent trial was obtained from Dubinsky Papers, box 319, folder 12B. See, for example, *Birds of a Feather*, leaflet produced by the ILGWU as the trial approached.
44. See "The People v. Benedicto Macri," transcript of trial proceedings, 5780/170, Kheel Center Archives, box 1. Also M. Epstein, *Jewish Labor in the U.S.A.* (Hoboken, N.J.: KTAV, 1969), 308–9.
45. Maxine Lurye to David Dubinsky, Mar. 18, 1952, Dubinsky Papers, box 297, folder 2A.
46. Anna Lurye to Gov. Thomas E. Dewey, Nov. 2, 1951, and Min Lurye Matheson to District Attorney Frank Hogan, n.d., ibid.
47. Donald G. Coe, American Broadcasting Company, to Joe Mazur, ILGWU Local 22, May 6, 1952, ibid.
48. Min L. Matheson to David Dubinsky, May 27, 1954, ibid.
49. S. Hoffman, oral history interview, Mar. 6, 1998, NPOHP, tape 2, side 1.

4. Building a Union Infrastructure

1. B. Greenberg, oral history interview, Jan. 11, 1995, Northeastern Pennsylvania Oral History Project, University of Wisconsin–Stevens Point (hereafter NPOHP), tape 1, side 1.
2. J. Silverman, oral history interview, July 16, 1993, NPOHP, tape 2, side 2.
3. Copies of *Needlepoint* were accessed at the Wyoming Valley District Office library, formerly located at 37 So. Washington St., Wilkes-Barre. The newsletters, kept in a filing cabinet in the library, were not indexed. Permission to copy was granted by Lois Hartel.
4. For examples of the variety of information contained in *Needlepoint*, see the issues of June 1954, May 1956, September 1958, April 1959, September 1962, and February 1963.
5. *Needlepoint*, February 1963, 1.
6. G. Price, "The Sweatshop" (1911), in *Out of the Sweatshop*, ed. L. Stein (New York: Quadrangle, 1977), 182–84.

Notes

7. ILGWU, *Report and Proceedings of the Tenth Convention of the* ILGWU (New York, 1910) and *Report and Proceedings of the Nineteenth Convention of the* ILGWU (New York, 1925).

8. J. Schereschewsky, "Stress and Strain" (1915), in Stein, *Out of the Sweatshop*, 186–87.

9. L. Johnson, "The First of Its Kind," in Stein, *Out of the Sweatshop*, 187–89. For further discussion of the history of the Joint Board of Sanitary Control and the ILGWU's health care centers, see ILGWU, *News History: The Story of the Ladies' Garment Workers* (New York, 1950); G. Tyler, *Look for the Union Label* (Armonk, N.Y.: M. E. Sharpe, 1995), esp. chap. 9.

10. "Area Garment Workers to Open Health Center," *Scranton Times*, June 2, 1948, A9. See also ILGWU, *Report of the General Executive Board to the Twenty-seventh Convention of the* ILGWU (New York: ABCO Press, 1950), 123.

11. M. Matheson, oral history interview, Dec. 5, 1988, NPOHP, tape 1, side 2. Min also discussed the health care center in an interview with the historian Sheldon Spear. See S. Spear, *Wyoming Valley History Revisited* (Shavertown, Pa.: Jemags, 1994).

12. *Needlepoint*, February 1963. See also the issues of April 1956, December 1958, and May 1961.

13. *Needlepoint*, December 1959.

14. Statistics on Health Care Center services up to and including 1953 are from "Union Health Group Gave 123,000 Tests in Penna. Districts," *Justice*, July 15, 1954, 14. Statistics on services through 1958 are from *Needlepoint*, December 1958.

15. *Needlepoint*, February 1960.

16. On *Pins and Needles* and the ILGWU's Labor Stage, see H. Broun, *"Pins and Needles,"* in Stein, *Out of the Sweatshop*, 249–51. Also see ILGWU, *News History; Report of the Educational Department,* ILGWU, June 1, 1944–Dec. 31, 1946 (New York, 1947), and *Report of the General Executive Board to the Twenty-fourth Convention of the* ILGWU, May 27–June 8, 1940 (New York: ABCO Press, 1940), 142–44; Tyler, *Look for the Union Label*, chap. 15.

17. Clementine Lyons, oral history interviews, July 5, 1990, NPOHP, tape 1, side 1, and July 24, 1993, ibid., tape 1, side 2. A slight distinction existed between the Wyoming Valley Chorus and the Northeast Department Chorus. For large-scale performances, such as ILGWU conventions, members of other ILGWU districts within the Northeast Department would join with Wyoming Valley members. Thus, on these occasions, the chorus was more appropriately referred to as the Northeast Department Chorus, or simply the ILGWU Chorus.

18. Program for *My Name Is Mary Brown: A Musical Narrative Presented by The Northeast Department,* ILGWU, *1950*. Our thanks to Helen Weiss for giving us the program.

19. On Unity House, see K. Wolensky, "Unity House: A Worker's Shangri-La," *Pennsylvania Heritage,* Summer 1998, 20–29.

20. *Needlepoint*, February 1960.

21. *Needlepoint*, May 1952.

22. *Needlepoint*, April 1959. On the Knox Mine Disaster, see R. Wolensky and K. Wolensky, "Disaster—Or Murder in the Mines?" *Pennsylvania Heritage,* Spring 1998, 4–11; and R. Wolensky, K. Wolensky, and N. Wolensky, *The Knox Mine Disaster: The Final Years of the Northern Anthracite Industry and the Effort to Rebuild a Regional Economy* (Harrisburg: Pennsylvania Historical and Museum Commission, 1999), and *Final Breach: The Knox Mine Disaster, the Contract-Leasing System, and the Demise of the Northern Anthracite Industry* (Urbana: University of Illinois Press, forthcoming).

23. *Needlepoint*, May 1962.

24. "ILGWU Chorus Sings at Stevenson Rally," *Wilkes-Barre Record*, Sept. 14, 1956, 8.

25. The lyrics of the ILGWU songs were provided by Clementine Lyons, a founding member.

26. There are a few variations of "You Gotta Know the Score." This version, introduced in the early 1960s, was edited from an earlier version that referred to Stevenson and Kefauver. Later versions mentioned Hubert Humphrey and Robert F. Kennedy. See ILGWU, *Northeast Department Anniversary, Thirty-second Convention of the ILGWU,* May 17, 1965 (New York: Astoria Press, 1965). In the 1970s and 1980s the song was again amended to refer to George McGovern, Jimmy Carter, and Walter Mondale.

27. Gingold is quoted in H. Crone, *35 Northeast: A Short History of the Northeast Department of the ILGWU, 1935–1970* (New York: ILGWU, 1970).

28. *Needlepoint,* December 1957.

29. *Needlepoint,* May 1955, January 1956, September 1958, June 1960. On Unity House, see K. Wolensky, "Unity House: A Worker's Shangri-La"; and on the political performances of the chorus, including its participation in an October 1960 rally on Public Square for John F. Kennedy, see ILGWU, *Silver Jubilee—Wyoming Valley District Council, Northeast Department* (New York, 1962).

30. There are several variations of "Look for the Union Label." The source for this version is ILGWU, *Northeast Department Anniversary.*

31. *The Northeast Sings,* recording (Northeast Department, ILGWU, New York, 1965); ILGWU, *Report of the General Executive Board to the Thirty-second Convention of the ILGWU* (New York: ABCO Press, 1965), and *Anniversary* (program for the Northeast Department's anniversary, 1935–65, presented at the Thirty-second Convention of the ILGWU, May 17, 1965, Miami Beach).

32. William Gable, oral history interview, Dec. 3, 1999, NPOHP, tape 1, side 1.

33. "500 at Testimonial Hail Min Matheson in Wilkes-Barre," *Justice,* Dec. 15, 1953, 11.

5. Constructing an Activist Union

1. On the ILGWU's workers' education initiatives, see M. Danish, *The World of David Dubinsky* (Cleveland: World, 1957); D. Dubinsky and A. Raskin, *David Dubinsky: A Life with Labor* (New York: Simon & Schuster, 1977); ILGWU, *News History: The Story of the Ladies' Garment Workers* (New York, 1950); and G. Tyler, *Look for the Union Label* (Armonk, N.Y.: M. E. Sharpe, 1995). Also the ILGWU's convention reports from 1918 to 1995 routinely discuss union educational activities; for early discussions see, for example, ILGWU, *Report and Proceedings of the Fourteenth Convention of the ILGWU* (1918); *Report and Proceedings of the Fifteenth Convention of the ILGWU* (1920); and *Report and Proceedings of the Sixteenth Convention of the ILGWU* (1922). The Education Department published occasional reports to discuss its history and activities, particularly during the 1930s and 1940s. A good example is *Report of the Educational Department of the ILGWU, June 1, 1944, to December 31, 1946* (New York: ILGWU, 1946). Finally, important archival sources include ILGWU Education Department Records, Kheel Center for Labor Management Documentation and Archives, School of Industrial and Labor Relations, Cornell University (hereafter Kheel Center Archives), 5780/049; the papers of Fannia Cohn, housed at the New York Public Library, Manuscripts and Archives Division (see, for example, Container 11, with information on lectures, panel discussions, and other ILGWU workers' education programs from the 1920s to the 1960s); and the papers of Mark Starr, housed at the Robert F. Wagner Archives, Tamiment Library, New York University.

2. M. Starr, "Workers' Education," *Harvard Education Review* 21, no. 4 (Fall 1951). A copy is in Fannia M. Cohn Papers, container 12, ILGWU, 1920–1960s, New York Public Library.

3. C. Sweeney, "The ILGWU Workers' University" (1920), in *Out of the Sweatshop*, ed. L. Stein (New York: Quadrangle, 1977), 245–48.

4. On the history of Unity House, see K. Wolensky, "Unity House: A Workers' Shangri-La," *Pennsylvania Heritage*, Summer 1998, 20–29.

5. Promotional brochure on Unity House, 1927, in Frederick F. Umhay Papers, Kheel Center Archives, 5780/005, box 5, folder 8.

6. Information on education programs at Unity House and related promotional material are in Education Department Records—ILGWU, Kheel Center Archives, 5780/049, box 3, folder 20; box 41, folders 18–20; and box 42, folders 1–8. Numerous ILGWU convention reports discuss Unity House and its education programs. See, for example, *Report of the General Executive Board to the Nineteenth Convention of the ILGWU*, May 7–17, 1928 (New York: ILGWU, 1928), 286–87; *Report of the General Executive Board to the Twenty-third Convention of the ILGWU*, May 3, 1937 (New York: ABCO Press, 1937), 165; and *Report of the General Executive Board to the Twenty-fifth Convention of the ILGWU*, May 29–June 9, 1944 (New York: ABCO Press, 1944), 165–66. See also *Unity News*, Aug. 24, Aug. 26, and Aug. 29, 1935, ILGWU Research Department Records, Kheel Center Archives, 5780-056, box 13, folder 9. Finally, Fannia Cohn discusses Unity House education programs in numerous writings. In the Fannia M. Cohn Papers, see, for example, "Educational Department of the ILGWU: An Analysis," container 9, and "Workers' Education and the Garment Workers," *New Leader*, Nov. 12, 1927 (n.p.), container 12.

7. T. Mathews and R. Hostetter, oral history interview, July 3, 1993, Northeastern Pennsylvania Oral History Project, University of Wisconsin–Stevens Point (hereafter NPOHP), tape 1, side 2.

8. Ibid.

9. M. Berger, oral history interview, Dec. 16, 1997, NPOHP, tape 1, side 1.

10. Unity House brochure, 1939, in Gus Tyler Papers, Kheel Center Archives, 5780/052, box 25, folder 16; ILGWU, *Report of the General Executive Board to the Twenty-third Convention of the ILGWU*, May 3, 1937 (New York: ABCO Press, 1937), 165.

11. V. Reisel, "Garment Workers' Country Club," *Survey Graphic*, September 1948, 388–91.

12. Eleanor Roosevelt, "Unity House Gives Youth Chance to Enjoy Country life," *New York World Telegram*, reprinted in ILGWU, *News History*, 109.

13. On the Union Counselor Program and the ILGWU's role in the Community Chest, see *Needlepoint*, July 1953, May 1961; and "37 Complete Community Chest's Union Counseling Course," *Wilkes-Barre Record*, Feb. 10, 1948, 4; "Need for Housing Stressed at Union Training Program," *Wilkes-Barre Record*, May 20, 1949, 3; "Area Residents Enrolled in ILGWU Health, Welfare Counselor Training Course," *Pittston Sunday Dispatch*, Feb. 19, 1956, 2; and "'Oscars' Awarded ILGers by 'Chest' in Wilkes-Barre," *Justice*, Feb. 1, 1954, 8.

14. *Needlepoint*, July 1953.

15. W. Matheson to ILGWU Education Dept., Nov. 18, 1953, in Wyoming Valley District library, Wilkes-Barre.

16. *Needlepoint*, December 1953, June 1954, June 1956; also see "Officers' Training Class Is Graduated by Pittston Local 295, Garment Workers," *Wilkes-Barre Record*, Feb. 24, 1957, 8.

17. M. Matheson, oral history interview, Dec. 5, 1988, NPOHP, tape 1, side 2.

18. *Needlepoint*, May 1955.

19. *Justice*, August 1954, 3.

20. On Dan Flood, see Kashatus, "'Dapper Dan' Flood: Pennsylvania's Legendary Congressman," *Pennsylvania Heritage*, Summer 1995, 4–11.

21. *Needlepoint*, August 1957 and September 1957.

22. *Needlepoint,* September 1957 and September 1962.

23. *Needlepoint* consistently reported on the district's education programs. See, for example, the issues of March 1957, September 1958, December 1959, May 1961, and August 1961.

24. C. Lyons, oral history interview, July 24, 1993, NPOHP, tape 1, side 1.

25. *Needlepoint,* July 1953.

26. *Needlepoint,* September 1957.

27. *Needlepoint,* September 1962.

28. R. Reuter, oral history interview, Mar. 6, 1998, NPOHP, tape 1, side 2.

29. B. Greenberg, oral history interview, Jan. 11, 1995, NPOHP, tape 1, side 2.

30. *Needlepoint,* August 1961.

31. J. Justin, oral history interview, Jan. 28, 1988, tape 1, side 2.

32. "Mayor Proclaims Nov. 16 as Garment Workers' Day in Pittston," *Pittston Sunday Dispatch,* Nov. 10, 1957, 38.

33. "Clothing Workers' Unions to Merge," *Harrisburg Patriot-News,* Feb. 14, 1995, B6. Also, on the ILGWU's political activism, see, for example, Danish, *World of David Dubinsky;* Dubinsky and Raskin, *David Dubinsky;* ILGWU, *News History;* and Tyler, *Look for the Union Label.*

34. *Needlepoint,* January 1956.

35. Ibid.

36. *Needlepoint,* March 1956.

37. *Needlepoint,* September 1958.

38. *Needlepoint,* September 1962.

39. Ibid.

40. M. Matheson, oral history interview, June 28,1990, NPOHP, tape 1, side 1.

41. S. Bianco, oral history interview, Mar. 11, 1996, NPOHP, tape 1, side 1.

42. *Needlepoint,* September 1962.

43. *Needlepoint,* May 1961.

44. *Needlepoint,* September 1958.

45. *Needlepoint,* March 1952.

46. *Needlepoint,* September 1958, September 1962.

47. On deindustrialization in the anthracite region, see T. Dublin, *When the Mines Closed* (Ithaca, N.Y.: Cornell University Press, 1998); D. Miller and R. Sharpless, *The Kingdom of Coal: Work, Enterprise, and Ethnic Communities in the Mine Fields* (Philadelphia: University of Pennsylvania Press, 1985); and R. Wolensky, K. Wolensky, and N. Wolensky, *The Knox Mine Disaster: The Final Years of the Northern Anthracite Industry and the Effort to Rebuild a Regional Economy* (Harrisburg: Pennsylvania Historical and Museum Commission, 1999), and *Final Breach: The Knox Mine Disaster, the Contract-Leasing System, and the Demise of the Northern Anthracite Industry* (Urbana: University of Illinois Press, forthcoming).

48. Kashatus, "'Dapper Dan' Flood," 7.

49. *Needlepoint,* January 1956 and March 1957.

50. *Needlepoint,* September 1958.

51. W. Shelton in *AFL-CIO News,* August 1958, 4.

52. *Needlepoint,* June 1960.

53. For further information on Congressman Flood, his influence on the federal budget, and his stature in the local community and in Washington, see Kashatus, "'Dapper Dan' Flood"; Miller and Sharpless, *Kingdom of Coal,* esp. chap. 9 and Epilogue; and G. Crille, "The Best Congressman," *Harper's,* January 1975, 11–16.

54. *Needlepoint,* June 1960. Even after Min and Bill Matheson departed the Wyoming Valley in 1963, Flood could count on continued support from the ILGWU's 10,000-plus local members. Discussions of the union's support for his federal initiatives after the enactment of the Area Redevelopment Act and before Min and Bill's departure may be found in the following issues of *Needlepoint:* June 1960, May 1961, September 1962, and February 1963. Interviews conducted as part of NPOHP confirmed the ILGWU's continued support of Flood to the end of his congressional career in 1980. See, for example, interviews with Sam Bianco, Mar. 11, 1996; Congressman Flood, July 5, 1990; Lois Hartel, June 17, 1993, and Mar. 16, 1996; and Sol Hoffman and Ralph Reuter, Mar. 6, 1998.

55. S. Bianco, oral history interview, Mar. 11, 1996, tape 1, side 2.

56. Commonwealth of Pennsylvania, *The Pennsylvania Manual* (Harrisburg, 1956).

57. During the twentieth century the governorship of Pennsylvania was held by a Republican for 72 years, by a Democrat for 28 years. In addition to George Earle (1935–39) and George Leader (1955–59), Democrats included David Lawrence (1959–63), Milton Shapp (1971–79), and Robert P. Casey (1987–95).

58. M. Matheson, oral history interview, Nov. 30, 1982, tape 2, side 1.

59. G. Leader, personal communication, Aug. 12, 1994.

60. G. Leader, oral history interview, May 30, 1995, NPOHP, tape 1, side 1.

61. Ibid.

62. Ibid.

63. Ibid., side 2.

64. For a history of Governor Leader's administration, see R. Cooper and R. Crary, *The Politics of Progress: Governor Leader's Administration, 1955–1959* (Harrisburg: Penns Valley Publishers, 1982); also see *Report of Governor George M. Leader to the General Assembly of the Commonwealth of Pennsylvania* (Harrisburg, 1959).

65. Min L. Matheson to ILGWU PEP Club participants, Oct. 27, 1955, memorandum, George M. Leader Papers, MG 207, Pennsylvania State Archives, General Correspondence File, box 49, folder 6.

66. *Needlepoint,* May 1955.

67. G. Leader, oral history interview, May 30, 1995, tape 1, side 1.

68. Ibid., side 2.

69. Min L. Matheson to Thomas Hodges, Governor's Office, July, 5, 1955, and announcement regarding Governor Leader's participation in WILK radio program, July 11, 1955, Leader Papers, box 49, folder 6.

70. *Needlepoint,* January 1956.

71. Commonwealth of Pennsylvania, *The Pennsylvania Industrial Development Authority: An Assessment* (Harrisburg, 1979); T. Dublin, "Attracting Business to the Anthracite Region, 1940–1970: Promises and Performance," paper presented to the Hagley Research Seminar, Wilmington, Del., May 1997; S. Spear, *Wyoming Valley History Revisited* (Shavertown, Pa.: Jemags, 1994).

72. For further information on these and related statistics, see Commonwealth of Pennsylvania, *Pennsylvania Industrial Development Authority;* Dublin, "Attracting Business to the Anthracite Region"; Spear, *Wyoming Valley History Revisited.* Data on unemployment in Luzerne County are derived from Pennsylvania Department of Labor and Industry, Bureau of Employment Security, *Labor Market Letter, Wilkes-Barre/Hazleton Area,* November 1968 and December 1970. Data on unemployment in Lackawanna County are derived from Pennsylvania Department of Labor and Industry, Bureau of Employment Security, *Labor Market Letter, Scranton and Lackawanna County,* November 1968 and December 1970.

73. Spear, *Wyoming Valley History Revisited*, 235.
74. On the Unity House theater and its dedication, see K. Wolensky, "Unity House: A Workers' Shangri-La"; program for the dedication of the Unity House Theater, June 2, 1956, provided by Clementine Lyons; and ILGWU, *Report of the General Executive Board to the Twenty-ninth Convention of the ILGWU*, May 1, 1956 (New York: ABCO Press), 203–4.
75. On other reforms proposed by the Leader administration, see Cooper and Crary, *Politics of Progress*. Governor Leader discusses ILGWU support for his initiatives in his oral history interview, May 30, 1995, tape 1, sides 1 and 2. The ILGWU discusses its support for Leader and his initiatives in numerous issues of *Needlepoint*, including ,January 1956, May 1956, September 1958, and February 1959. Also see Min L. Matheson to Governor Leader, Mar. 20 and Aug. 15, 1957, and Leader to Matheson, Jan. 17 and Aug. 20, 1957, Leader Papers, box 49, folder 6.
76. *Needlepoint*, February 1960.
77. "County Was Once Solid Republican Territory," *Wilkes-Barre Times-Leader*, Oct. 7, 1983, 2A, 10A; Commonwealth of Pennsylvania, *Pennsylvania Manual*. In one example of the realignment of voter registration patterns in Pennsylvania steel towns, voters in a heavily industrialized blue-collar ethnic area of Allegheny County overturned decades of Republican dominance by overwhelmingly supporting Democrats during the height of the New Deal era. See, for example, Eric Leif Davin, "Blue Collar Democracy: Class War and Political Revolution in Western Pennsylvania, 1932–1937," *Pennsylvania History* 67, no. 2 (Spring 2000): 240–97.
78. "PA Dress Strikers Are New Pioneers, Dubinsky Tells Workmen's Circle," *Justice*, May 15, 1958, 10.

6. Our Demands Must Be Met: The 1958 General Dress Strike

1. "Pittston Dress Makers Reassured by Union," *Wilkes-Barre Record*, May 2, 1957. On Antonini see P. Cannistraro, "Luigi Antonini and the Italian Anti-Fascist Movement in the United States, 1940–1943," *Journal of American Ethnic History* 5 (Fall 1985): 21–40; and R. Fillippelli, "Luigi Antonini, the Italian-American Labor Council, and Cold War Politics in Italy, 1943–1949," *Labor History* 33, no. 1 (1992): 102–25.
2. ILGWU, *Report of the General Executive Board to the Thirtieth Convention of the ILGWU*, May 1, 1959 (New York: ABCO Press, 1959), 88–93. On the 1958 strike also see D. Melman, "The Cause and Effect of the ILGWU Dress Industry General Strike of 1958," master's thesis, Baruch College, City University of New York, and New York State School of Industrial and Labor Relations, Cornell University, 1994.
3. U.S. Department of Labor, Bureau of Labor Statistics, *1955 Earnings in the Dress Manufacturing Industry* (Washington, D.C.: U.S. Government Printing Office, 1956), 7.
4. See Melman, "Cause and Effect of the ILGWU Dress Industry General Strike," esp. 14–16.
5. Ibid., 21–23.
6. "Penn. Dress Rallies Hear Hochman Sound Clarion Call for Solidarity," *Justice*, Feb. 15, 1958, 4.
7. "Dress Workers to Conduct Parley," *Scranton Tribune*, Feb. 20, 1958, 2; "Dress Strike Imminent as Parleys Fail," *Women's Wear Daily*, Feb. 21, 1958, 2; "ILGWU Calls Meeting to Prepare for Strike," *Scranton Times*, Feb. 26, 1958, 2; "ILGWU Officials Meet in Bethlehem to Form Needed Strike Committees," *Scranton Times*, Feb. 27, 1958, 3.
8. "ILGWU Set for Walkout: Wage Negotiations Break Down," *New York Times*, Feb. 27, 1958, 2.

9. "Dress Talks Continuing: No Strike This Weekend, Dubinsky Says," *Wilkes-Barre Times-Leader Evening News*, Feb. 28, 1958, 3.

10. "105,000 in 7-State Dress Strike," *Justice*, Mar. 1, 1958, 2.

11. "Strike of Garment Workers Called: Wage Parley Ends Without Agreement," *Scrantonian*, Mar. 2, 1958, 1.

12. "Garment Strike Looms at Friday Midnight," *Wilkes-Barre Times-Leader Evening News*, Feb. 24, 1958, 2.

13. "Avert Garment Strike," *Wilkes-Barre Times-Leader Evening News*, Feb. 26, 1958, 26.

14. Ibid.

15. "Dress Plant Strike Looms for Tuesday," *Wilkes-Barre Times-Leader Evening News*, Mar. 3, 1958, 3.

16. "Strike Tomorrow at 115 Local Dress Shops," *Wilkes-Barre Times-Leader Evening News*, Mar. 4, 1958, 3.

17. Ibid.; "Dress Workers' Strike Slated for Tomorrow," *Scranton Times*, Mar. 4, 1958, 3.

18. "Picture of a Union: The ILGWU," *New York Times*, May 27, 1959, sec. 10. The ILGWU estimated that as of the end of 1958 there were 16,000 women's and children's clothing factories in the United States employing 575,000 workers, 440,000 of whom were organized by the ILGWU. Annual payroll totaled $1.5 billion and wholesale volume of business was $6.3 billion annually.

19. "Dress Makers Strike in 7 States," *Wilkes-Barre Times-Leader Evening News*, Mar. 5, 1958, 34; "Dress Strike Called 100% Effective," ibid., Mar. 6, 1958, 1.

20. "Ladies' Dress Output to Be Halted," *Scranton Tribune*, Mar. 5, 1958, 1.

21. "Peace Bids Speeded in Dress Halt," *Women's Wear Daily*, Mar. 6, 1958, 1.

22. "Dress Shops Struck," *Wilkes-Barre Times-Leader Evening News*, Mar. 5, 1958, 3.

23. Julius Hochman, manager of ILGWU's Joint Dress Board, to Harry Schindler, ILGWU Scranton district manager, telegram, Mar. 6, 1958, in ILGWU Wyoming Valley District library, Wilkes-Barre.

24. "Pickets Walk into Three Dress Shops to Halt Operations," *Wilkes-Barre Times-Leader Evening News*, Mar. 6, 1958, 3; "Dress Strike Called 100% Effective," ibid., 1.

25. "Efforts Pushed to End Dress Workers Strike: Early Accord Seen," *Scranton Times*, Mar. 6, 1958; "N.Y. Meeting Scheduled in Dress Strike," *Wilkes-Barre Times-Leader Evening News*, Mar. 7, 1958, 3; "End of Dress Strike Sought by Mediators," *Wilkes-Barre Times-Leader Evening News*, Mar. 8, 1958, 3.

26. ILGWU, *Report of the General Executive Board to the Thirtieth Convention of the ILGWU*, May 11, 1959 (New York, 1959), 138–43. Wage increases were estimated to be $75 million over the three-year contract term, the bulk of which would be made up by consumers through increased retail prices. Though the ILGWU had explored the idea of a union label over 35 years earlier, 1958 was the first time that the union succeeded in negotiating the union label into an industry-wide contract. The label was intended to boost sales of apparel made in union shops while simultaneously reducing the market share of apparel made in nonunion firms. The union label became one of the hallmarks of the ILGWU and was advertised as the "Symbol of Decency, Fair Labor Standards, and the American Way of Life."

27. "3-Year Pact Reached in New York Dress Strike," *Wilkes-Barre Times-Leader Evening News*, Mar. 11, 1958, 1; Melman, "Cause and Effect of the ILGWU Dress Industry General Strike," 26–27.

28. "3-Year Pact Reached in New York Dress Strike," *Wilkes-Barre Times-Leader Evening News*, Mar. 11, 1958, 1.

29. "Garment Strikers Pay Visit to Mercy Hospital," ibid., Mar. 12, 1958, 30.

30. "Union to Pay Benefits if Strike Continues," ibid., 3.
31. "Local Dress Strike Talks Are Sought," ibid., Mar. 13, 1958, 3; "40,000 More Back on Job," ibid., 1.
32. Ibid.
33. Ibid.; "Local Dress Talks Today in New York," *Wilkes-Barre Times-Leader Evening News,* Mar. 14, 1958, 3.
34. "Dress Strike Settlement Hinted," ibid., Mar. 17, 1958, 1.
35. "Local Strikers Are Told Pennies Are Issue," ibid., Mar. 20, 1958, 3; "6 More Shops Resign from PGMA," ibid., Mar. 25, 1958, 13.
36. "Some Plants Leave PGMA, Work Monday," ibid., Mar. 21, 1958, 3; "6 More Shops Resign from PGMA," ibid., Mar. 25, 1958, 13.
37. "Strike Still On against Pennsylvania Holdouts," *Justice,* Apr. 1, 1958, 2.
38. "PGMA Reports Progress in Strike Issue," *Wilkes-Barre Times-Leader Evening News,* Mar. 31, 1958, 3; "PGMA Rejects Union Terms to End Strike," ibid., Apr. 1, 1958, 3.
39. "Woman Picket Is Punched at Wyoming," ibid., Mar. 13, 1958, 1; "Local Dress Talks Today in New York," ibid., Mar. 14, 1958, 3.
40. "Mrs. Matheson and 3 Others Arrested in Pittston Strike," ibid., Apr. 1, 1958, 3.
41. "Two Trucks Damaged by Women Strikers," ibid., Apr. 2, 1958, 3; "Women Pickets Block Attempt to Remove Dresses from Plant," *Scranton Times,* Apr. 11, 1958, 1.
42. "Red Paint Splashed in Dress Strike Here," *Wilkes-Barre Times-Leader Evening News,* Apr. 9, 1958, 3.
43. Proclamation by Sheriff Joseph Mock of Luzerne County, published in *Wilkes-Barre Record,* Apr. 9, 1958, in David Dubinsky Papers, Kheel Center for Labor Management Documentation and Archives, School of Industrial and Labor Relations, Cornell University (hereafter Kheel Center Archives), 5780/002, box 297, folder 4E.
44. "New Election Asked for by PGMA Workers," *Wilkes-Barre Times-Leader Evening News,* Apr. 3, 1958, 3; "PGMA Lawyer Says Union Fears Vote," ibid., Mar. 26, 1958, 36; "ILGWU Pushes for Solution without PGMA," ibid., Apr. 4, 1958, 3.
45. "ILGWU Pushes for Solution without PGMA," ibid., Apr. 4, 1958, 3.
46. Ibid.
47. "Dress Holdouts Bar Pact 3rd Time," *New York Times,* Apr. 2, 1958, 2.
48 "Bitterness Marks Meeting That Fails to End Dress Strike," *Wilkes-Barre Times-Leader Evening News,* Apr. 7, 1958, 3.
49. "Dubinsky Tells 1,000 Cheering Workers Here: 'Hoods' Block Dress Pact," *Scranton Times,* Apr. 15, 1958, 1; "Leader Calls Meeting in Dress Strike," *Scranton Tribune,* Apr. 15, 1958, 2.
50. "Win Dress Strike in Pennsylvania as 'Invisible Forces' Are Defeated," *Justice,* Apr. 16, 1958, 2.
51. ILGWU, *Report of the General Executive Board to the Thirtieth Convention,* 138–43; *Report of the General Executive Board to the Thirty-first Convention of the ILGWU,* May 23, 1962 (New York, 1962), 102–4.
52. "Dubinsky States 'Strike Is Over,'" *Scranton Times,* Apr. 18, 1958, 18.
53. "Jenkins Owners Fail in Attempts to Open Padlocks in Pittston," *Justice,* Aug. 1, 1959, 8; "Indict 9 in Pittston for Bosses' Beating of Jenkins Strikers," ibid., Sept. 15, 1959, 10.
54. Min L. Matheson, district manager, Wyoming Valley District, to William Batt, Jr., secretary of labor and industry, Apr. 10, 1959, Wyoming Valley District library, Wilkes-Barre.
55. "Area Garment Industry Is Being Probed," *Scranton Times,* Apr. 11, 1958, 2; "End 2-Year Jenkins Halt Thru Full N'east Pact," *Justice,* Jan. 14, 1960, 4

56. "Takers of 'Fifth' Due to Be Quizzed Again," *Scranton Tribune*, May 13, 1959, 1, 18; "450 in Pa. Picket Bitter Holdouts at 8 Dress Shops," *Justice*, July 15, 1958, 2. In an untaped interview on Nov. 30, 1988, Min Matheson discussed an unprecedented meeting between her and mobsters to resolve the conflict in her district. In the spring of 1958 she was invited to New York by Abe Chait, an underworld figure with ties to the garment industry. Although she did not inform ILGWU headquarters of the invitation, she accepted with some misgivings. Upon entering the meeting place at Seventh Avenue and West 36th Street, she found Russell Bufalino, Dominick Alaimo, and Angelo Sciandra sitting with Chait and others. Shocked and perturbed at the presence of the three, she told Chait, "If I had known that you were going to have these rotten people here, these people from the sewers, I wouldn't have come. They are worse than from the sewers and I won't have anything to do with them!" She then stormed out of the building. One of Chait's lieutenants followed her, grabbed her by the arm, and informed her that she'd "better go back in." Mrs. Matheson reluctantly returned to the meeting and discussed the situation with Chait and the others. Her failure to inform ILGWU headquarters about the meeting caused her some anxiety; "If D.D. [Dubinsky] ever knew that I went to New York for such a meeting, he would have killed me!" Matheson's apparent boldness demonstrated her strong desire to settle a contentious labor situation. Her concern over Dubinsky also indicated some of the tension that existed between headquarters and her district.

57. On the McClellan Committee's investigation of organized crime's influence in the apparel industry in Pennsylvania see *The McClellan Committee Hearings* (Washington, D.C.: Bureau of National Affairs, 1958); Select Committee on Improper Activities in the Labor or Management Field, *Hearings*, 86th Cong., 1st sess., June 30–July 3, 1958; and S. Petro, *Power Unlimited: The Corruption of Union Leadership–A Report on the McClellan Committee Hearings* (New York: Ronald Press, 1959). On the Bufalino organized crime family and its affiliates with business interests in northeastern Pennsylvania, see the prepared statement of Malcolm L. Lazin, chairman, Pennsylvania Crime Commission, before the U.S. Senate Permanent Subcommittee on Investigations, Feb. 23, 1983 (Washington, D.C.: U.S. Government Printing Office, 1983).

58. "450 in Pa. Picket Bitter Holdouts at 8 Dress Shops," *Justice*, July 15, 1958, 2; "Dubinsky Battles Dress Holdouts," *New York Times*, Apr. 16, 1958, 4. A serious conflict developed between the ILGWU in New York and the Wyoming Valley District shortly after the general dress strike of 1958 was settled but while negotiations with the PGMA were still ongoing. The conflict provides important insight regarding internal union politics and the relationship between the international and one of its larger districts. According to Mrs. Matheson:

> Right after the '58 strike [was settled] they pulled us out again. They told us to [continue to] strike. It was to suit New York's purposes and to disrupt our situation here. But if I failed to carry out the order, which was the decision of the General Executive Board, you know, there's discipline in the union. I was furious. I called Dubinsky. We had the people out for a few days.
>
> So, although I was of New York and from New York and the New York leaders were first my friends, if you know what I mean, originally I used to get pretty fed up with their nonsense. Right after these few days of striking, I knew the pay envelopes would be depleted and I was very, very unhappy about the whole action. Although the girls stayed out, I was just hoping they would say to me, "The hell with you," and work, that's how mad I was.

It was a Thursday in the evening and we were just closing up shop and my superior [probably David Gingold] called me up and he was in a rage. He must have just come out of a meeting where he likely had a big fight with the New York guys. And he said, "If he [Dubinsky] wants a strike, let them pay for it. Let 'em pay for it. I want you to pay twenty dollars to every striker for strike benefits and I'll make him pay it back to you." I said, "We have a lot of people out." "Pay it. I want you to do it right away so they can't stop you." It's Thursday, like four, five o'clock in the afternoon, and there were a couple of girls in the office who heard this conversation because he was screaming so loud they could hear it.

So I called the bank presidents and I said I need fifty thousand dollars; we have to cover our shops. The girls were out and we've got to give them twenty dollars so they'll have it to buy groceries. The banks cooperated. We had accounts in a couple of banks and we got the money and Friday we passed it out. The girls were very surprised, you know. And when they started writing letters, the little devils, to Dubinsky thanking him for the strike benefit, we were surprised! He called me up and he said he was surprised! And what right did I have to pay this strike benefit without authorization? Right away I knew there was big trouble. I could've answered and said [my superior] told me to do it, but I didn't 'cause I realized instantly that he was in deep, dark trouble.

I don't know why I did it. I can't even explain to you to this day. I never told Dubinsky the truth. You know when it bothered me? When D.D. [Dubinsky] died. That day I couldn't rest. Why didn't I ever sit down with D.D. and tell him the truth? I don't know what would've happened. It couldn't be good for the department, and it couldn't be good for the politics of the union.

I went in to New York and he [Dubinsky] was walking around talking in Jewish, Russian, and English and screaming, "Fifty thousand dollars, fifty thousand dollars you gave away without permission. What made you do it?" He couldn't believe it. I said [to myself], that's the end of the strikes in Pennsylvania. . . . They came in and took my treasury away. They told all the bank presidents that they were not to honor my signature or my checks. The girls who were in the office and knew why I did it wanted me to tell the truth. I said, "No, kids, it's gonna be worse. We can live with this. We'll keep right on doing what we're doing. They want to sign the checks, good. It's a pleasure. I don't have to waste my time signing checks. . . . Let the bastards sign it. Who cares?" So they took away our treasury. (Oral history interview, Nov. 31, 1982, Northeastern Pennsylvania Oral History Project, University of Wisconsin–Stevens Point, tape 1, side 2)

59. See Melman, "Cause and Effect of the ILGWU Dress Industry General Strike," 30–33.
60. "Min L. Matheson New ILGWU Union Label Dept. Head," ILGWU news release, Jan. 23, 1963, in Dubinsky Papers, box 218, folder 5.

7. Importing Apparel and Exporting Jobs

1. "Impressive Tribute to Min L. Matheson," *Pittston Sunday Dispatch*, Jan. 27, 1963, 6. There are several versions of exactly why Min was relocated to New York. In an untaped discussion (Sept. 24, 1999), Sam Bianco explained that it was common (though unofficial) knowl-

edge at the time that Dubinsky removed Min from her post as district manager because he had come to view her as too powerful, influential, and outspoken. Allegedly she was removed to New York to curtail her power. Betty Greenberg, Min's daughter, has shared the same sentiment with us. Ralph Reuter provided two rationales for Min's departure (untaped discussion, June 15, 2000). First, despite some gains in the 1958 strike, wages in Wyoming Valley dress factories remained lower than those in New York's dress industry and thus the region continued to provide an outlet to jobbers looking for less expensive contractors and thereby threatened the New York market. Min was viewed as at least partly responsible for the lack of wage parity. Second, her longtime association with the more leftist wing of the ILGWU and Charles Zimmerman in particular was viewed as out of step with the more conservative trade unionist philosophy that had come to dominate the ILGWU leadership. The second point was reiterated by John Justin in an oral history interview on June 6, 2000. Justin opined that Sol "Chick" Chaikin, Northeast Department assistant director and later president of the ILGWU (1975–86), purged key department leaders whom he considered out of step with the union's philosophy at the time. Min alluded to several of these issues that had generated internal conflict between the international union and the Wyoming Valley District:

> See, there were a lot of internal fights in the union. Between the New York union and the Pennsylvania union there was always bad blood. Maybe I shouldn't even tell you about this; I rarely ever talk about it. Of course, the shops ran away to get cheaper labor. But New York is a little like the Russians: what we do you have to do. Whether it fits or doesn't fit, we still have to do it. And we're always suspect because we're working outside of the big [city].
>
> You see, the big part of the union was in New York. And we always had to apologize! While we had the harder times and we were weaker and all the time we're fighting and working like hell and building a union and trying to build up conditions. But we were always suspect because the wages weren't good enough, the conditions weren't good enough. But that certainly wasn't because we weren't trying but because I'd been chairlady of Local 22 [in New York] and they always expected the impossible.
>
> Now we had a lot of these big affairs because we wanted to show 'em how much support we had from the people in this community. And instead of being in our favor, they were against us. "Look how they're growing. She wants to run the [whole] union. Look how big she's getting." We were getting too many members. I resented it very much and they would do things to make it impossible for us to get work out of New York. Also for their petty factional reasons, [I thought] maybe the shops would die. (Oral history interview, Nov. 31, 1982, Northeastern Pennsylvania Oral History Project, University of Wisconsin–Stevens Point [hereafter NPOHP], tape 1, side 2)

Whatever the reason for her departure from the Wyoming Valley, directing the Union Label Department was an important and highly visible job—precisely the type of job that demanded a leader with a public presence and commitment to the labor movement.

2. "Mrs. Matheson Key Figure in Area," *Wilkes-Barre Times Leader*, Feb. 4, 1963, 2A.

3. M. Matheson, oral history interview, Dec. 5, 1988, NPOHP, tape 2, side 1.

4. Ibid. On the Agnes disaster and recovery, see R. Wolensky, *"Better than Ever":The Flood Recovery Task Force and the 1972 Agnes Disaster* (Stevens Point, Wis.: UWSP Foundation Press, 1993).

5. "Min L. Matheson Dies at 83," *Wilkes-Barre Times Leader*, Dec. 11, 1992, 2D.

6. "Min Matheson: The Lady and the Gangsters," *Justice*, Mar. 1, 1993, 6.

7. J. Williams, oral history interview, July 31, 1984, NPOHP, tape 1, side 1.

8. "Min Matheson Made Difference for the Better," *Wilkes-Barre Citizens' Voice*, Dec. 9, 1992, 20.

9. "Plaque Honoring Min Matheson Unveiled," ibid., Sept. 25, 1999, 1, 3.

10. J. Mazur, oral history interview, Mar. 6, 1998, NPOHP, tape 1, side 1.

11. On the Lattimer Massacre, see M. Novak, *The Guns of Lattimer* (New York: Basic Books, 1978), and K. Wolensky, *The Lattimer Massacre*, Historic Pennsylvania Leaflet no. 15 (Harrisburg: Pennsylvania Historical and Museum Commission, 1997).

12. On deindustrialization, see B. Bluestone and B. Harrison, *The Deindustrialization of America* (New York: Basic Books, 1994).

13. G. Tyler, *Look for the Union Label* (Armonk, N.Y.: M. E. Sharpe, 1995), 278.

14. The ILGWU first reported on the increase in imported Japanese scarves in *Report of the General Executive Board to the Twenty-ninth Convention of the ILGWU*, May 10, 1956 (New York: ABCO Press, 1956). Other sources that discuss the history and impact of imports include S. Chaikin, *A Labor Viewpoint: Another Opinion* (Monroe, N.Y.: Library Research Associates, 1980); Tyler, *Look for the Union Label*; M. Wark, "Fashion as a Culture Industry," in *No Sweat: Fashion, Free Trade, and the Rights of Garment Workers*, ed. A. Ross, 227–48 (New York: Verso, 1997).

15. For statistics on apparel workers' wages, see T. Bowman, "Cheap Labor Blamed for Plant Failure," *Harrisburg Patriot-News*, Nov. 19, 1996, B3, B6; Chaikin, *A Labor Viewpoint*, esp. 8 and 86; ILGWU, *Report of the General Executive Board to the Thirty-fifth Convention of the ILGWU*, May 31, 1974 (New York: ABCO Press, 1974), and *General Executive Board Report to the Thirty-eighth Convention of the ILGWU*, May 27–June 3, 1983 (New York: ABCO Press, 1983); Commonwealth of Pennsylvania, *The Pennsylvania Garment Industry: Foreign Competition Costing Garment Workers' Jobs* (Harrisburg, 1985); and Pennsylvania Department of Labor and Industry, Bureau of Employment Security, *Pennsylvania Employment and Earnings* 21, no. 1 (Harrisburg, January 1975).

16. Chaikin, *A Labor Viewpoint*; ILGWU, *Report of the General Executive Board to the Thirty-fifth Convention*.

17. Tyler, *Look for the Union Label*, 288.

18. Chaikin, *A Labor Viewpoint*, 23.

19. For further information on various apparel trade arrangements and loopholes, see S. Chaikin, "The Needed Repeal of Item 807.00 of the Tariff Schedules of the United States," testimony presented to the Subcommittee on Trade, Committee on Ways and Means, U.S. House of Representatives, Mar. 25, 1976; Chaikin, *A Labor Viewpoint*, esp. chaps. 1, 6, and 7; Commonwealth of Pennsylvania, *Pennsylvania Garment Industry*; ILGWU, *Report of the General Executive Board to the Thirty-fifth Convention*; *General Executive Board Report and Record, Thirty-sixth Convention of the ILGWU*, May 27, 1977 (New York: ABCO Press, 1977); *General Executive Board Report and Record of Proceedings, Thirty-seventh Convention of the ILGWU*, Sept. 28–Oct. 3, 1980 (New York: ABCO Press, 1980); Tyler, *Look for the Union Label*, esp. chaps. 22 and 23; Wark, "Fashion as a Culture Industry."

20. Tyler, *Look for the Union Label*, 267–68.

21. "U.S. and Mexican Organizers Meet with Highlander in Mexico," *Highlander Reports*, January–May 1998, 1; "Third World Nations Join Outcry over Exploitation in Marianas," *Label Letter* (AFL-CIO), May–June 1998, 1; "Corporate Responsibility: Sweatshops—Morality's Role in World Economy Divides Labor Leader, Educator," *Harrisburg Patriot-News*, Mar. 29, 1998, D1, D10.

261

Notes

22. For a further discussion of this point, see Tyler, *Look for the Union Label*, 296; Alan Howard, "Labor, History, and Sweatshops in the New Global Economy," in Ross, *No Sweat*, 151–72; and Carl Proper, "New York: Defending the Union Contract," in Ross, *No Sweat*, 173–92.

23. For statistics on the growth of imports, see Chaikin, *A Labor Viewpoint*, 32; Commonwealth of Pennsylvania, *Pennsylvania Garment Industry*, 4–5; "Don't Let Another U.S. Industry Be Destroyed," *Wilkes-Barre Citizens' Voice*, Oct. 24, 1984, 14; ILGWU, *Report of the General Executive Board to the Thirty-fifth Convention*, 53, and *General Executive Board Report to the Forty-second Convention of the ILGWU*, June 1995 (New York, 1995), 33. The statement that 88% of garments sold in the United States in 2000 were manufactured overseas is derived from an address by Edgar Romney, executive vice president of UNITE!, to members of the Pennsylvania, Ohio, and South Jersey Joint Board, Tamiment Resort, Bushkill, Pa., June 11, 2000.

24. Historical summaries of ILGWU membership and industry employment trends are included in union convention reports. See, for example, *General Executive Board Report to the Forty-second Convention*, esp. 1–40. Also see C. Kernaghan, "Paying to Lose Our Jobs," in Ross, *No Sweat*, 79–94; "Fighting to Survive: ILGWU in 2-way Squeeze," *New York Post*, Mar. 14, 1977, 14; "Ailing Garment Workers' Union Faces More Problems," *New York Times*, Nov. 29, 1981, A6.

25. Data on 2000 apparel industry employment and related projections are provided courtesy of the Pennsylvania Department of Labor and Industry, Center for Workforce Information and Analysis, Harrisburg.

26. "Garment Makers Suffer in Sales of Imported Clothes," *Philadelphia Evening Bulletin*, June 24, 1977, 34.

27. L. Hartel, oral history interview, Mar. 16, 1996, NPOHP, tape 1, side 1.

28. Ibid.

29. Ibid., side 2; *Pennsylvania Manufacturing Trends, 1988 to 1992* (Harrisburg: Pennsylvania State Data Center, September 1993).

30. Data on apparel-related employers are derived from "Units and Employees in the Apparel Industry," a report prepared for us by the Pennsylvania Department of Labor and Industry, Center for Workforce Information and Analysis, 2000.

31. L. Gutstein, oral history interview, June 26, 1997, NPOHP, tape 1, side 2.

32. "Leslie Fay: A Pattern for Life?" *Wilkes-Barre Times Leader*, Apr. 3, 1994, 2B.

33. "Arbitrator Rules against Leslie Fay Co.," *Wilkes-Barre Citizens' Voice*, Oct. 28, 1993, 5; "Leslie Fay Goes Astray," *Justice*, October 1993, 1; "Campaign to Save Leslie Fay Jobs: Which Side Are You On?" *Leslie Fay Workers' Newsletter—ILGWU*, November 1993, 1; "Audit Report Details Fraud at Leslie Fay," *Wall Street Journal*, Mar. 28, 1995, B1.

34. "Leslie Fay Wants to Cut U.S. Thread," *Wilkes-Barre Times-Leader*, Mar. 31, 1994, 1A, 14A; "Leslie Fay Aims to Shut Down All U.S. Plants," *Women's Wear Daily*, Mar. 31, 1994, 1, 3; "Leslie Fay: Rosy Despite Losses," *Scranton Tribune*, Mar. 31, 1994, B11.

35. "Unionized Leslie Fay Employees Protest Cutbacks in Work Week," *Wilkes-Barre Times-Leader*, Mar. 1, 1994, 1A, 10A; "Furloughed Garment Workers Stage Protest at Leslie Fay," *Wilkes-Barre Citizens' Voice*, Mar. 1, 1994, 3, 33; "Leslie Fay Employees Keep Job Vigil, *Wilkes-Barre Times-Leader*, Mar. 2, 1994, 3A. Correspondence on the unfair labor practice complaint, provided by Lois Hartel of UNITE!, consists of a copy of the complaint and cover letter filed by the law firm of Handler, Gerber, Johnston, and Aronson of Camp Hill, Pa., dated Apr. 6, 1994. On the Northeastern Pennsylvania Stakeholders Alliance see "Area Leaders Back Union Cry to Keep Leslie Fay in the USA," *Wilkes-Barre Times-Leader*, Mar. 18, 1994, 14A. Various internal memoranda also provided by Hartel describe the role of the Alliance as well as agendas, dates, times, and places of meetings.

36. Letters provided by Lois Hartel from Bishop James C. Timlin, Mar. 10, 1994; Bishop Harold Weiss, Mar. 16, 1994; and William George, Mar. 17, 1994, express support for the work of the Alliance and voice opposition to Leslie Fay's plans to eliminate production in Pennsylvania.

37. The first quote is derived from an ILGWU poster headed "Leslie Fay Gives Workers Rotten Eggs for Easter," inviting workers to a meeting on Apr. 6, 1994, to discuss the situation, provided by Lois Hartel. The second quote is from an advertisement placed by the ILGWU in the *New York Times*, May 1, 1994, A8.

38. "Local Garment Workers Battle Foreign Competition," *Wilkes-Barre Times-Leader*, Mar. 24, 1994, 1F, 2F. On the union's protest at the home of Michael Babcock, see "Garment Worker Arrested in Protest," *Greenwich Times*, May 15, 1994, 3; "Leslie Fay Workers Take Protest to Connecticut," *Wilkes-Barre Citizens' Voice*, May 15, 1994, 6.

39. Gov. Robert P. Casey to Michael Babcock, May 17, 1994; Lt. Gov. Mark S. Singel, news release, May 17, 1994; copies of both provided by Lois Hartel.

40. "ILGWU Strikes Leslie Fay," *Wilkes-Barre Citizens' Voice*, June 1, 1994, 2; "Leslie Fay, Union Draw Battle Lines," *Wilkes-Barre Times-Leader*, June 2, 1994, 1A, 18A; "Clash of the Titans," *Wilkes-Barre Citizens' Voice*, June 2, 1994, 1, 3.

41. Midwest Center for Labor Research, *Social Cost Analysis of Layoffs of Leslie Fay Workers in Pennsylvania* (Chicago, January 1994). On Grossman's comments and Wofford's visit see "Leslie Fay Closing Will Produce Ripples," *Wilkes-Barre Times-Leader*, June 5, 1994, 3A; "Sen. Wofford Offers Support to ILGWU Strikers," *Wilkes-Barre Citizens' Voice*, June 5, 1994, 3, 18.

42. "Leslie Fay a No-Show at Fact-Finding Session," *Wilkes-Barre Citizens' Voice*, June 8, 1994, 5, 9; "Leslie Fay Plight Examined," *Wilkes-Barre Times-Leader*, June 8, 1994, 1A, 12A; "A Sampling of Opinions Heard during Hearing," *Scranton Tribune*, June 8, 1994, A1, A8.

43. L. Hartel, oral history interview, Mar. 16, 1996, tape 1, side 2.

44. "Kanjorski Plan 2d Leslie Fay Hearing, Seeks Appearance by John Pomerantz," *Women's Wear Daily*, June 15, 1994, 2, 21.

45. "AFL-CIO to Call for Leslie Fay Boycott," *Wilkes-Barre Times-Leader*, June 9, 1994, 3A; "Union Marches in Big Apple; Leslie Fay Issues Statement," *Wilkes-Barre Citizens' Voice*, June 10, 1994, 4; "Leslie Fay Responds to Claim," *Scranton Tribune*, June 10, 1994, A1, A13; "Leslie Fay Workers Supported by Catholics," *Catholic Light*, June 23, 1994, 10.

46. "Leslie Fay Locks Four Struck Plants," *Wilkes-Barre Times-Leader*, June 16, 1994, 2A; "Tearing at Fabric of Women's Lives," *Philadelphia Inquirer*, June 17, 1994, B1, B9; "Leslie Fay, Union Report No Progress in Talks," *Wilkes-Barre Time-Leader*, June 18, 1994, 2A; "Leslie Fay Need Not Hold Jobs, Says Judge," *Wilkes-Barre Times-Leader*, June 22, 1994, 1A, 12A.

47. General Assembly of Pennsylvania, House Resolution no. 351, June 20, 1994; "Leslie Fay Workers Supported by Catholics"; Jay Mazur's comments appeared in "As We Fight for Jobs, We're Offered Cynicism," *Wall Street Journal*, letter to the editor, June 24, 1994, D2.

48. "Striking Leslie Fay Workers Protest Big Raises for Bosses," *New York Post*, June 30, 1994, 4; "Renowned Labor Mediator to Hear Leslie Fay Dispute," *Wilkes-Barre Citizens' Voice*, June 29, 1994, 3; "Leslie Fay to Give Bonuses to Brass," *Wilkes-Barre Times-Leader*, July 1, 1994, 1A, 16A.

49. "A Settlement Is Reached to End Leslie Fay Strike," *New York Post*, July 12, 1994, 6; "ILGWU, Leslie Fay Praise Tentative Labor Agreement," *Wilkes-Barre Citizens' Voice*, July 12, 1994, 1; "Leslie Fay Union Approves Job-Cut Pact," *Wilkes-Barre Times-Leader*, July 14, 1998, 1A, 14A.

50. Jay Mazur is quoted in "ILGWU, Leslie Fay Praise Tentative Labor Agreement." Though

263

Notes

optimistic, the strike organizer Tom Mathews cautioned about the unfinished business of the agreement in "Leslie Fay Deal Remains Murky," *Wilkes-Barre Times-Leader,* July 13, 1994, 1A, 10A. On the prayer service and picnic, see "Service Offers Thanks for End of Leslie Fay Strike," *Wilkes-Barre Citizens' Voice,* July 15, 1994, 1A, 2A.

51. As of 2000, much of Leslie Fay's apparel manufacturing was contracted to Doall Industries, Inc., in the San Marcos Free Trade Zone, El Salvador.

52. "Leslie Fay to Sell Some of Its Businesses as It Operates under Chapter 11," *Wall Street Journal,* Mar. 20, 1995, B6; "Leslie Fay to Close Its Last U.S. Factory," *New York Times,* May 8, 1995, D3; "Dressmaker to Close Facility in Pennsylvania, *Wall Street Journal,* May 8, 1995, B8.

53. In June 1997 Leslie Fay emerged from Chapter 11 and reported $203 million in sales for the first six months of the year. The company continued to restructure by selling off or eliminating production of unprofitable apparel. Virtually all of its production originated in Guatemala. See "Bankruptcy Court Confirms Plan for Reorganization," *Wall Street Journal,* Apr. 22, 1997, B4; "Earnings Nearly Doubled; Reorganization Costs Drop," ibid., May 27, 1997, A4.

54. Telephone conversation with Pearl Novak, Sept. 25, 1998. For Novak's perspective on the Leslie Fay strike, see also her oral history interview, June 19, 1997, NPOHP, tape 1, side 2.

Epilogue

1. Both the domestic and international focus of the garment workers' union and its educational mission are often discussed in its newsletter, UNITE! See, for example, the issue of September–October 1997. which discusses several of the union's education initiatives. Also see October 1995, which discusses a major campaign targeting the retailer GAP and its exploitation of overseas workers. Another excellent source of information on education initiatives is union convention reports. See, for example, *General Executive Board Report to the Forty-second Convention of the ILGWU,* June 1995, the sections "Education" and "Health and Safety." This report also discusses the union's international activities, as does S. Chaikin, *A Labor Viewpoint: Another Opinion* (Monroe, N.Y.: Library Research Associates, 1980); and G. Tyler, *Look for the Union Label* (Armonk, N.Y.: M. E. Sharpe, 1995), esp. chaps. 22 and 23.

2. Several chapters in A. Ross, ed., *No Sweat: Fashion, Free Trade, and the Rights of Garment Workers* (New York: Verso, 1997), discuss the involvement of the ILGWU/UNITE! in domestic and international industry organizations. See, for example, A. Ross, "After the Year of the Sweatshop: Postscript," 291–97. On the Apparel Industry Partnership, see "Groups Reach Agreement for Curtailing Sweatshops," *New York Times,* Nov. 5, 1998, A20.

3. See, for example, "Major Retailers Still Avoid Responsibility," UNITE!, October 1995, 3; "No More Sweatshops," UNITE!, September–October 1997, 12; and "Walking the Line on Abuse," *Los Angeles Times,* July 25, 1995, D1. On the Fashion Trendsetters List, see "No Sweat Fashion," in Ross, *No Sweat,* 298.

4. U.S. Department of Labor, *Protecting America's Garment Workers: A Monitoring Guide* (Washington, D.C.: U.S. Government Printing Office, 1998). And see "Major Retailers Still Avoid Responsibility," UNITE!, October 1995, 3; "No More Sweatshops," UNITE!, September–October 1997, 12; Tyler, *Look for the Union Label,* chaps. 22 and 23.

5. For a review of the exhibit, see G. Palladino, "Between a Rock and a Hard Place: A History of American Sweatshops, 1820–Present," *Public Historian* (Smithsonian Institution) 21, no. 1 (Winter 1999): 143–46. On the El Monte raid, see J. Su, "El Monte Thai Garment Workers: Slave Sweatshops," in Ross, *No Sweat,* 143–50.

6. See "Don't Let Another U.S. Industry Be Destroyed," *Wilkes-Barre Citizens' Voice*, Oct. 24, 1984, 14; "It's Time to Stem the Tide of Imports of Apparel," ibid., Mar. 18, 1995, 12.

7. R. Hostetter, oral history interview, July 3, 1993, Northeastern Pennsylvania Oral History Project, University of Wisconsin–Stevens Point (hereafter NPOHP), tape 1, side 1.

8. See ILGWU, *General Executive Board Report to the Forty-second Convention*, esp. the section "The Fight for Free Trade," 29. Also see "Members Mobilize to Defeat NAFTA Fast Track," *UNITE!*, September–October 1997, 5.

9. L. Hartel, oral history interview, Mar. 16, 1996, NPOHP, tape 1, side 2.

10. "The U.S. Garment Industry: Is It Coming Apart at the Seams?" *Philadelphia Inquirer*, Dec. 11, 1994, E1, E6.

11. Hartel, oral history interview, Mar. 16, 1996, tape 1, side 2.

12. "Don't Let Another U.S. Industry Be Destroyed," *Wilkes-Barre Citizens' Voice*, Oct. 24, 1984, 14.

13. Text provided by *UNITE!* Research Department, 1998.

14. Hartel, oral history interview, Mar. 16, 1996, tape 1, side 2.

15. "Clothing Workers' Unions to Merge," *Harrisburg Patriot-News*, Feb. 15, 1995, B6. Also see "Union Merger Is Announced," *New York Times*, Feb. 21, 1995, C7; "Two Big Apparel Unions to Be UNITE!(d)," *New York Times*, Feb. 26, 1995, E2; and "For Better or Worsted—An Extra Large Proposal," *Washington Post*, Feb. 18, 1995, D1. On the $10 million organizing campaign see "New Needle-Trades Union Pledges $10 Million to Organize Workers," *Wall Street Journal*, Feb. 28, 1995, A1.

16. "Organizing: Keeping Our Promise," *UNITE!*, October 1995, 3.

17. "Union Takes 'Sweatshop' Protest to Hecht's Store," *Washington Post*, Dec. 21, 1996, D2; "2 Unions to Seek Board Representation at K-mart," *New York Times*, Mar. 28, 1996, C1; "2 Unions Sue K-mart over Shareholder Vote," *New York Times*, June 13, 1996, D12.

18. On the Stop Sweatshops Act, see "A 'United' Message: Eliminate Sweatshops," *Wilkes-Barre Citizens' Voice*, Aug. 2, 1998, 7, 10; and *Congressional Record*, 105th Cong., 1st sess., Jan. 7, 1997, E71. As of this writing, the legislation had not yet been introduced in the 107th Congress.

19. Protests at the WTO Seattle conference drew widespread media coverage and public attention during late November and early December 1999. See, for example, "Protestors Could Steal the Show at Seattle Trade Talks," *New York Times*, Nov. 29, 1999, A1; "Session Disrupted, Trade Ministers Insist They Will Continue" and "National Guard Is Called to Quell Trade Talk Protests," ibid., Dec. 1, 1999, A1, A2; and "UNITE! Rallies against World Trade Organization," *Wilkes-Barre Citizens' Voice*, Dec. 1, 1999, 3. The China Trade Bill likewise drew wide media coverage and public interest. See, for example, "Organized Labor Says China Backers Will Pay a Price," *Wall Street Journal*, May 25, 2000, 1.

20. "Sweatshop Issue Escalates with Sit-Ins and Policy Shifts," *Chronicle of Higher Education*, Mar. 10, 2000, 38–40.

About the Authors

Kenneth C. Wolensky is a historian at the Pennsylvania Historical and Museum Commission, Harrisburg. His research focuses on labor, working-class, and public policy history. He holds a doctoral degree in adult education from The Pennsylvania State University, a master's degree in public policy from the University of Delaware, and a B.A. in history from College Misericordia. He is co-director of the Northeastern Pennsylvania Oral History Project, serves on the board of the Pennsylvania Labor History Society, and is a lecturer for the Pennsylvania Humanities Council and The Pennsylvania State University.

Nicole Wolensky is a graduate of Marquette University, Milwaukee, with a bachelor's degree in sociology and psychology, and won the Knudten Award for excellence in sociology. She is enrolled in the graduate program in sociology at the University of Iowa, specializing in social psychology and conducting research on former Soviet republics.

Robert P. Wolensky is Professor of Sociology and co-director of the Center for the Small City at the University of Wisconsin–Stevens Point. He is co-director of the Northeastern Pennsylvania Oral History Project and has researched and written about various aspects of the history and culture of the anthracite region, including coal mining and the flood occasioned by Tropical Storm Agnes in 1972. He has served as a scholar-in-residence at the Pennsylvania Historical and Museum Commission and as Visiting Fellow at the Institute for Research in the Humanities, University of Wisconsin-Madison. He holds M.A. and Ph.D. degrees in sociology from The Pennsylvania State University and an A.B. degree in sociology from Villanova University.

Kenneth, Nicole, and Robert Wolensky are co-authors of *The Knox Mine Disaster: The Final Years of the Northern Anthracite Industry and the Effort to Rebuild a Regional Economy,* published by the Pennsylvania Historical and Museum Commission in 1999; and *Final Breach: The Knox Mine Disaster, the Contract-Leasing System, and the Demise of the Northern Anthracite Industry,* forthcoming from University of Illinois Press.

Index

Gable, William, 99, 100, 105, 110, 117, 236
garment industry
  Canada, 197
  Carribean, 206
  central Pennsylvania, 11
  China, 197, 206, 232
  Coney Island, 17
  contractors (contracting), 13, 23–24,
    16–17, 25–29, 27, 34, 163–64, 178,
    195–97, 210, 223, 241
  Cotton Textile Agreement, 200
  Dallas/Fort Worth, 223
  Dominican Republic, 206
  effects of immigration on, 12–13, 222, 223
  El Salvador, 215
  employment in Pennsylvania, 11, 33–38,
    40–41, 156–57, 187–89, 202–3
  Guatemala, 201, 208, 209, 210, 215
  Haiti, 197
  homeworking, 7, 16, 223
  Hong Kong, 197, 198, 206
  "inside" system of production, 13, 16
  Japan, 196–97, 199–200
  jobbers, 16, 24, 34, 42, 164, 178, 196–97,
    241
  Korea, 197, 206
  Lehigh Valley, 11
  Lower East Side, 12,14,15, 17, 206, 233
  Manhattan, 3, 12–13, 34, 223
  manufacturer/jobber-contractor system,
    16, 24, 241
  Mexico, 198, 201
  migration within United States, 8–9, 24,
    34–38, 196, 233
  migration to overseas nations, 195–207,
    220–22, 226, 233
  Multi-Fiber Agreement (MFA), 200
  Northern Mariana Islands, 201
  "outside" system of production, 16, 196
  Philadelphia, 13–15, 20, 34, 44, 202,
  "runaway" factories, 3, 23–25, 102, 104,
    196
  retailers, 197–98, 202, 211, 212, 224, 230
  "section-work" system, 38, 163
  Singapore, 197
  South Korea, 197
  Sri Lanka, 201
  Taiwan, 198, 206

"transshipping," 200
U.S. import statistics, 199–200, 202
Vietnam, 201
wages and compensation, 28, 38, 41, 163,
  179, 179–84, 197–98, 201, 209, 211,
  223–24
Western Pennsylvania, 12, 27
whole-garment system, 38
See also anthracite region; International
  Ladies' Garment Workers' Union;
  Leslie Fay, Inc.; New York, New York;
  Wyoming Valley District–ILGWU
General Agreement on Tariffs and Trade
  (GATT), 226
Genovese (organized crime family), 53
George, William (Bill), 209
Gifford, Kathy Lee, 230
Gingold, David
  migration of garment industry to Penn-
    sylvania, 39–41
  oral history, 6, 41, 236
  organizing in Pennsylvania, 42–45, 68,
    114
  role in 1958 General Dress Strike, 164,
    171, 172, 174, 177
  writings, 6, 39–41
Glassberg, Abraham, 118, 172, 177
Glen Lyon Manufacturing Company, 79
Gluken Bra Company, 79
Gotham Knife Cutters' Association of New
  York and Vicinity, 13
Goode, Mayor Wilson, 132
"Great Revolt," 17
Greater Wilkes-Barre Chamber of Com-
  merce, 118
Greenberg, Betty (Matheson), 50–51, 52,
  62–63, 73, 75, 90–91, 132, 191, 236
Greenberg, Larry, 236
Grossman, Howard, 210
Guisto, John, 84–85
Gustin, Alfred, 131
Gutstein, Leo, 35, 36–37, 57–58, 205–6,
  236

Hanover Industrial Park, 207
Harrisburg, 35, 42
Harrisburg Patriot-News, 229
Harsey Blouse Company, 80